Paul Biber's Experiences at Szentandrás Airship Base Near Timisoara

Code Name "ADEBAR" 1915 and 1916

Dr. Jörg Biber

Translated by Catherine Lawler

Acknowledgements
Color aircraft profiles © Bob Pearson. Purchase his CD of WWI aircraft profiles for $50 US/Canadian, 40€, or £30, airmail postage included, via Paypal to Bob at: **bpearson@kaien.net**

For our aviation books, please see our website at: **www.aeronautbooks.com**.

I am looking for photographs of the less well-known German aircraft of WWI to complete this series. For questions or to help with photographs you may contact me at **jherris@me.com**.

ISBN: **978-1-953201-29-4**

Design and layout: Jack Herris
Cover design: Aaron Weaver
Digital photo editing: Jack Herris

Aeronaut Books
Books for Enthusiasts by Enthusiasts
www.aeronautbooks.com

Table of Contents

Chapter 1: Preface

1.1 Book Concepts

Using photographs and corresponding notes taken from Paul Biber's war photo album, this book will outline his experiences at the Szentandrás airship base near Timisoara (Temesvár, present day Romania), Hungary during WWI. Furthermore, the utilization of the airship base from 1915–1918 and its temporary use of code name "ADEBAR" will be highlighted in greater detail.

Paul Biber captured many of the photographs himself, using his own camera and developed the images in order to mail them home as field-post cards. He had asked his father to keep the many photographs which frequently contained notes and personal experiences on the back. After the war, he organized the images in chronological order and added them to his war photo album, supplementing the album with notes.

My father, Paul Biber was born in Dresden on July 6, 1891; he passed away on September 1, 1957 in Glashütte when I was only six years old. Fortunately, I have kept many of my father's documents in order to better remember him. Going through his materials has challenged me at times but overall, it supported my eagerness in following in his footsteps. If he was still alive, he would surely have enjoyed my following his career path. I became a toolmaker, a metal profession such as my father's. I was an instructor and a vocational teacher following my graduation as a qualified engineer educationalist from TU Dresden. Later on, I completed my doctorates at TU Dresden where I was an instructor for vocational schoolteachers in metal technology and mechanical engineering for 41 years. Like my late father, I strived to publish extensively. Over the past years, inspired by the collection of his documents, I finally took the time to analyze more closely some facets of my father's life. Among other publications, it is how this book came into existence.

After attending elementary and citizen school, Paul Biber completed an apprenticeship as a precision engineer with Messwerkzeugfirma Gebr. Keilpart in Dresden from April 1906. He passed his final exams in May 1910 with distinction. In 1912, he joined Dresden-based Firma Georg Rosenmüller where he manufactured measuring devices for flow measurement as an engineer from October 1913. Among other devices, G. Rosenmüller manufactured aeronautical (aircraft and airships) wind gauges (anemometers and airspeed indicators). On August 22, 1914, Paul Biber was drafted for military service.

In order to properly file the extensive photographic materials containing his experiences within the military context, a compilation of experiences of Hauptmann Max Macher was included among a variety of other references (see Kustan & Prenz at https://scharekf.hu/zeppelin.html, Max Macher).

Part 3 of the compilation contains detailed facts about the Szentandrás airship base which were utilized in specific parts of this book. Max Macher was born on February 8, 1882 in Gutenstein (Kärnten). He graduated in Marburg and thereafter studied engineering for 7 semesters, followed by 2 semesters of philosophy. He completed Reserveoffizersprüfung in 1906, followed by Ergänzungsprüfung zum Berufsoffizier. On November 1, 1908 he was promoted to Leutnant. In 1910, he completed an airship course. Leutnant Max Macher earned Diplom des Österreichischen Aero-Club Nr. 12 für Lenkballonfahrten (Austrian Aero-Club Number 12, Diploma for Dirigible Balloon Operations) on October 8, 1911. On October 19, 1912 he received Diplom des Österreichischen Aero-Club Nr. 71 für Freiballonfahrten (Austrian Aero-Club Number 17, Diploma for Hot Air/Free Balloon Operations) (see Kustan & Penz – Part 1).

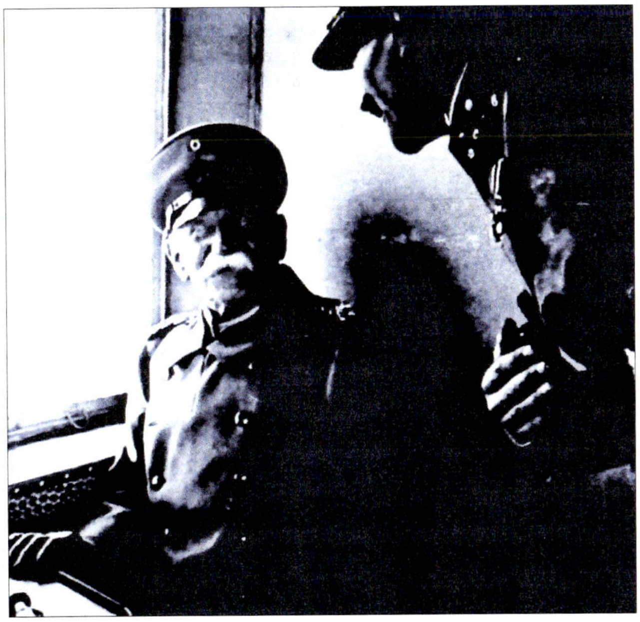

Image 0 Above: Graf Zeppelin and Hauptmann Macher onboard Navy airship "L 30" in 1916 (image: Schmeelke Archive).

Image 1 Above: LZ 17 "*Sachsen*". (Täger *Propellerblatt* Magazine, 2009)

Furthermore, accounts by Kornel Poppe (Kommandeur vom Luftschiffhafen Szentandrás) were utilized (see http://scharefk.hu/zeppelin.html, Kornel Poppe).

Kornel Poppe was born in Budapest on October 10, 1884. After one year of serving as a volunteer, he was promoted to Offizier in 1904. When WWI broke out, he served as Kommandant with a Luftwaffe Festungsballonabteilung (Air Force Balloon Department). Following his training at Luftschiffhafen Nordholz, he took over command of the newly established Luftschiffhafen in Timisoara (Hungary) in 1916 and 1917. On January 16, 1918 Kornel Poppe successfully completed a parachute jump. Oberstleutnant Kornel Poppe died in 1941.

1.2 Zeppelin Airships Before, and at the Beginning of WWI

Count Ferdinand von Zeppelin (1838–1917) had been considering the idea of constructing a rigid airship since 1873. Following his discharge from military service as Generalletunant in 1891, he focused all of his energy on making this idea a reality. The first design concept was completed in 1895 for which he was issued a corresponding patent. "When the government expert commission rejected the concept and subsequently its financing, Zeppelin utilized his entire family fortunes for the construction of his airship.

The first airship, LZ 1, rose into the sky on July 2, 1900, thereby proving the validity of his concept for a rigid airship." (see Bélafi, 1988, Cover)

"In October 1906, LZ 3 has two successful ascents from Lake Constance, conducts several maneuvers, and touches down safely on the water each time. As a result, Technische Hochschule Dresden (Dresden Technical University) issues the Count the honorary academic degree of "Doktor-Ingenieur Ehrenhalber" (Honorary Doctorate in Engineering Dr.-Ing. h.c., Dr.-Ing. E.h respectively.) as the first Scientific Institute in Germany (Bélafi und Schade, 2015) (siehe VDI 61.1917).

With the foundation of "Deutsche Luftschiffahrtsaktiengesellschaft" (German Airship Stock Company, Delag) on November 16, 1909 Count Ferdinand von Zeppelin established the world's first civil airship carrier. His goal was to build a network of civilian airship travel in order to demonstrate the peaceful utilization of airships. The implementation of his plans included the construction of airship hangars in various cities, including Düsseldorf, Hamburg, and Dresden. As a result, an initial foundation had been put in place for the foundation of a national airline network, a first in the history of aviation. Count Zeppelin deserves

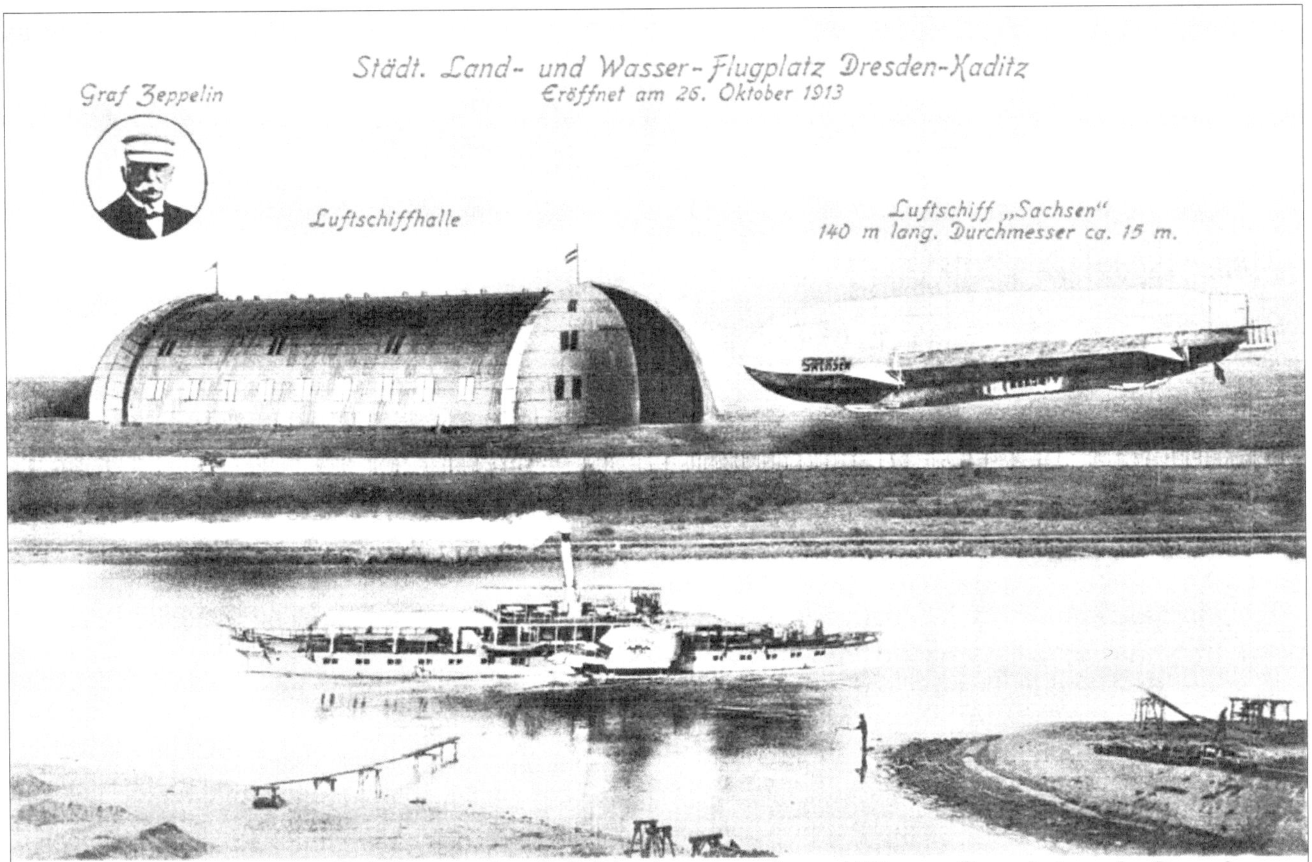

Eine aus Anlass der Eröffnung des Städtischen Land- und Wasserflugplatzes in Dresden-Kaditz publizierte Postkarte (Sammlung Täger)

Image 2 Above: Postcard published on the occasion of the inauguration of the city airport and seaport in Dresden-Kaditz (Täger-*Propellerblatt* Magazine, 2009)

credit for contributing to the peaceful utilization of airship aviation (see Bélafi, 1988, S. 104).

"On August 18, 1912, Zeppelin LZ 11 '*Victoria Luise*' landed in Dresden on the Elbe river for the first time. On October 28, the Dresden city council decided to establish an airship base and airport at Kaditz. (...) One year later, the king dedicated the world's first city airport and airship base. Within only six months, the world's largest airship hangar had been built on the Elbe river, measuring 192 meters in length, 58 meters in width, and 37 meters in height. The hangar could accommodate two Zeppelin airships" (Bélafi and Schade, 2015).

Delag received multiple Zeppelin airships in order to manage the many trips. From May 3, 1913, airship "*Sachsen*" (Saxony, LZ17) (see image 1) completed 419 trips in one year (managed by Delag), traveling 39,919 of kilometers while carrying 2,465 paying passengers. (see Bélafi, 1988, page 111)

On the afternoon of October 26, 1913 thousands of Dresden citizens flocked to Kaditz airport (the first city airport in the world) in order to attend the inauguration of the magnificent "Albert" airship hangar (see image 2). Zeppelin airship "*Sachsen*" LZ 17 was on display in front of the hangar; it had arrived from Leipzig the day before.

Following welcoming remarks issued by head mayor, Geheimer Rat Dr. Beutler to honorary guests including the king and his family, "*Sachsen*" airship departed on a one-hour trip toward Pirna. (see Fiedler, 1988, page 6). It was "*Sachsen's*" 200th trip at that time.

"In early 1914, LZ 17 "*Sachsen*" was tentatively used as a Navy training vessel in Dresden. In June of 1914, Count Zeppelin flew to Austria onboard "*Sachsen*", landing in Vienna in front of an enthusiastic crowd; it was his intent to demonstrate "that he wanted to commission his invention to peaceful air transport. Later, following the outbreak of the war, he said: "This war does not suit my plans. We intended to take airships out over the open seas and facilitate world travel. [Schillert, H.: Zeppelin. Bad Godesberg 1966, page 51]" (see Bélafi, 1988, page 111)

Delag was the first civilian air carrier in the world; it was quickly dissolved as a result of the

Image 3 Above: LZ 17 during its war time deployment on the Eastern Front (image: Racil Collection).

outbreak of the war.

Delag airships, "*Viktoria Luise*", "*Hansa*", and "*Sachsen*" were seized by the military; their purpose was to serve as training vessels for war airship aeronauts with immediate effect. (see Bélafi, 1988, page 111)

"At the outbreak of the war, the Army already had seven operational airships and the Navy had one. Army airship Z 10 was ready for immediate commissioning. Navy airship L 4 was still under construction. There were three older Delag airships, "*Sachsen*", "*Hansa*", and "*Viktoria Luise*". All three of them were remarkably successful passenger airships. It appears as though "*Sachsen*" and "*Viktoria Luise*" had been used as Navy training vessels while "*Hansa*" had been utilized by the Army. When the war broke out, four Army airships were stationed in the west and three in the east." (Marine, 1992/I, page 11)

"At the outbreak of the war, the military administration owned six war airships, five of which belonged to the Army and one to the Navy. In addition, the military had seized three airships formerly operated by Delag; these airships had been refitted to serve as military training vessels. In the course of the war, 90 war airships were built (from

Military Airship Distribution between the Army and the Navy During the War

Type of Airship	Army	Navy
Zeppelin	30	65
Schütte-Lanz	5	7
Parseval	1	3
Groß-Basenach	0	1

Bélafi, 1988, page 115

Image 4 Right: "Timisoara" airship base was established first, followed by "Jamboli" airship base in the "Balkan States" war region. (see Navy, I/1990, page 7)

a total of 119 ever built). This large number was possible because at that time, building a Zeppelin airship rarely took longer than four weeks." (Bélafi, 1988, page 114)

The General Staff viewed Zeppelin airships as an excellent means for both distant reconnaissance as well as bombing of targets far in the enemy Hinterland. The German General Staff – the Army, and particularly the Navy had inadequately developed their Air Force. They expected Zeppelin airships to perform miracles including the investigation and destruction of the war economy along with the demoralization of the population. (see Groehler 1985, page 71)

Count Zeppelin considered his Zeppelin airships to be suitable objects for the observation of enemy territories beyond the reach of enemy guns and sharpshooters. He thought of the commanding of troops from the air as a tactical scope of application for the military (see Bélafi, 1988, pages 112/113).

"From here on out, the construction of Zeppelin airships was part of the gigantic Imperial Armaments Program in Germany, resulting in vast amounts of capital flowing into the Zeppelin monopoly. (...) Within a very short amount of time, the Army and Navy had built more than 25 airship hangars, most of them in Western Germany. More airship hangars were established in German allied countries and later in occupied countries. (see Bélafi, 1988, page 114)

Zeppelin airship LZ 17 "*Sachsen*" was transferred from its home-base in Leipzig to Cologne at the outbreak of the war on August 1, 1914. "Kptlt. Lehmann, a former Navy officer and Delag employee was designated Kommandant. Kapitän Hacker, another former Delag employee was designated 2. Kommandant.

With armed forces onboard, "*Sachsen*" bombed Antwerp in August and September of 1914. The three attacks originated from a base in the Rhineland region, probably from Cologne." (see Marine, 1992/III, page 6)

Image 5 Above: Insertion of the substructure design into the excavation pit (see scharekf.hu)

Attacks by "LZ 17" (see image 3) took place on August 24, 1914, September 2, 1914, and September 22, 1914 (September 9 was an attempted attack only). It is possible that "LZ 17" was part of attacks taking place on October 2 and October 8, 1914. Thereafter, the airship was brought to Allenstein until February. In March and April, "LZ 17" participated in attacks on Bialystok and Wilna (for example on February 4, 1915) as well as in June 1915 on the Lomza train station. (see Stadtwiki Dresden)

"*Sachsen*" airship bombed Lomza on July 21/21. "Following a break due to a full moon, the airships deployed even more numerously in early August 1915. The following were bombed between 2 and 13. VIII: Bialystok railway station by "Sachsen" airship and Z. XII with a total of 2100 kg; Malkin railway station by Z. XII with 1650 kg, Sieldze and Nowo-Minsk railway stations whose utilization was particularly important to the Russians following the Warsaw occupation from 5. VIII. with 1600 kg by Z. XII., and 500 kg by L.Z. 39; Sztaroßjeltzy and Lapy railway junctions with 2000 kg by Z. XII; Wilna railway station by "Sachsen" with 600 kg; Nowo-Georgieest (surrounded since August 9) by L.Z. 39 with 500 kg.

The last service of "*Sachsen*" was in no way inferior to the ones of the past; its last attack on Wilna, where an ammunition train had apparently been hit while stopping on the platform, was a very important achievement. Thereafter, however, the airship was eliminated from war service and transferred to the Navy for training purposes." (Reprint, 1920, page 365)

Between May 1915 and May 1919, the Navy aeronautical department was relocated from Leipzig to Dresden. The department was tasked with the training of crews for new airships. Takeoff and landing support was provided by "III. Marine-Luftschifftrupp". From April 1915, the utilization of the new military airship hangar named "Johann" at Kaditz airfield was made available. The length of the hangar measured 184 meters.

"*Sachsen*" LZ 17 airship was used in part as a Navy training vessel in Dresden even from early

Image 6a Above: Attachment of prefabricated metal bracings to the hangar segments (see schrekf.hu)

Image 6b Above: Construction of barracks foundations to be utilized as warehouses or for crew housing. September 21, 1915. (image: Fabry Archive)

1915. In July 1915, Oberleutnant Ernst Scherzer took over command of LZ 17 "*Sachsen*" from Kapitänsleutnant Lehmann, operating the airship on the eastern front until September 1915 (see Ploes, 2018, page 17).

Thereafter, Oberleutnant Ernst Scherzer handed over "LZ 17" to the Navy for regular utilization as a training vessel. One year later, on September 6, 1916, LZ 17 "*Sachsen*" was disarmed in the town of Düren (North Rhine-Westphalia).

Following the withdrawal of the Navy aeronautical department in May 1916, airship hangars "*Albert*" and "*Johann*" in Dresden-Kaditz were utilized as reconstruction and repair facilities for airships in conjunction with large repair workshops. From December 1914 through January 1915, Z 11 was equipped with new engines, in January 1916 LZ 81 had its engines replaced at the same location.

Contracts through summer of 1917 pertained to the extension of various Army and Navy "P" type airships that were to be converted to "Q" type airships; some remaining contracts lasted through the end of 1917. (see Stadtwiki Dresden and Navy, 1992/IV, page 4)

Army airship aviation discontinued in the summer of 1917. As a result of available space at the Dresden-Kaditz airship base, Luftschiffer-Ersatzabteilung (LEA) Nr. 6 settled there. This department was responsible for the establishment and training of balloon units (Feldluftschiffer), commanded by Hauptmann Georg Bode. (see Stadtwiki Dresden)

Airships built by Zeppelin Luftschiffbau in Friedrichshafen and Potsdam were issued consecutive numbers. No airship was laid down without a number; the numbers were issued independently of the airship's delivery destination (Navy or Army). From January 1915, the shipyard numbers doubled as military designations of Army airships. From April 1915, the shipyard numbers were increased by 30 (shipyard number LZ 55–LZ 85), resulting in a gap of 30 numbers between LZ 39 and LZ 72.

"Since the beginning of the war, Friedrichshafen and Potsdam each delivered approximately the same number of airships. (...) Before the outbreak of the war, Friedrichshafen had built 25 airships." (Navy, 199/III, page 2)

Airships built in Friedrichshafen in August/September of 1915 were issued shipyard numbers 46, 47, 48, 49, 50, and 55. Of the Friedrichshafen airships, 46 (L14), 48 (L15), and 50 (L16) were transferred to the Navy while 47 (LZ 77) was transferred to the Army. 49 (LZ 79) and 55 (LZ 85) were transferred to the Army from Potsdam." (Navy, 1992/III, page 2)

During the first weeks of the war, four Zeppelin airships were shot down. "This caused the Prussian Army Command to issue a ban on airship missions during daylight hours. Enemy airship defense was particularly successful on Germany's western front." (Bélafi, 1988, page 116)

"The great losses of war airships on the western front forced German war strategists to move their airship operations to the east and to the Balkans, as enemy defense was considerably weaker there compared to the west. In addition, the German Army Command attempted to weaken the morale of the population by launching attacks on the enemy Hinterland." (see Bélafi, 1988, page 116)

Accordingly, within a very short period of

Image 7a Above: Raising of the hangar segments (see scharekf.hu)

Image 7b Above: Almost all segments for Timisoara airship hangar have been raised. The barracks which have been already constructed are visible in the background, October 21, 1915 (image: Fabry Archive).

Image 7c Above: The last hangar segment is raised at Timsoara, October 23, 1915 (image: Fabry Archive).

time, airship hangars were constructed in eastern countries allied with Germany, such as Jamboli in the Balkans, Szentandrás near Timisoara (see image 4), or Warsaw and Kowno. (see Bélafi, 1988, page 114)

1.3 September 1915 – Construction of an Airship Base with a Large Hangar Near Timisoara

From German military command's perspective, strategic reconnaissance and bombing were to be supported by airships as well. In order to quickly implement these changes, Chief of German General Staff, General von Falkenhayn requested the approval of k.u.k. Chief of General Staff, General Conrad von Hötzendorf for the construction of an airship storage hangar in southern Hungary on July 29, 1915. Thereupon, k.u.k. airship department ordered airship department Hauptmann (chief officer) Mannsbarth to press ahead with the construction of an airship hangar. It was quickly decided to establish the airship hangar and the operationally necessary buildings on a flat area approximately 2 km southeast of Szentandrás railway station in close proximity to the city of Timisoara.

The corresponding premises were leased (see Ploes, 2018, page 16) and an airship hangar was built on a large pasture extending between the towns of Szentandrás and Szentendre.

Prior to WWI, Timisoara was still part of Austria-Hungary and was located in southeastern Hungary on the border to Serbia and Romania. It was a lively trading town which culturally and economically blossomed in the second half of the 19th century. The main contributing factor had been the connection of Timisoara to the rail network (1857) and the canalization of Bega. The city was shaped by straight streets, large squares, and representative buildings. Timisoara was also known as "Little Vienna". In 1869, Timisoara was the second Hungarian city after Budapest to receive a horse streetcar. By late November of 1884, the first streetlights were operated with electrical power.

At the start of the war, approximately 75,000 people lived in Timisoara. At the time, Germany had the largest population.

In addition to its strategic location, the connection to the rail network as well as the availability of the necessary trade network were of high importance to the quick construction of an airship base in the vicinity of Timisoara.

Turkey had allied with the Central Powers; however, there was no direct land connection between Turkey and Austria-Hungary during the initial years of the war.

Mid 1915, Bulgaria tended to join the Central Powers, mobilizing its forces on September 23, 1915.

On October 5, 1915 British and French troops surprisingly landed in the Greek town of Thessaloniki (Saloniki for short) in order to attack Central Power troops (Austria-Hungary, Turkey, Germany, Bulgaria) via Serbia.

Romania had previously remained neutral; but its neutrality was not stable. Romania could easily switch sides and join the enemy if the opportunity arose. Therefore, it was imperative to act quickly (see Navy 1990/I, page 1).

As a result, new tasks were added to the Army's airship operations in the Balkans war zone. Airship hangars were to be constructed on short notice in Jamboli (Bulgaria) and Timisoara (southern Hungary).

In order to be able to utilize the Timisoara airship hangar in the course of the planned attack against Serbia, its construction was to be completed within 3–4 weeks. Therefore, hangar construction began right away, on September 3.

The airship hangar faced southwest, which was the prevailing wind direction at this location.

To meet the deadline, portable hangar segments made from prefabricated metal struts manufactured by B. Seibert company (Saarbrücken) were used by means of screw connections in order to insert the delivered foundation segments after the excavation of construction pits (see images 5 and 6a). Simultaneously, construction began for the foundations of the barracks buildings (see image 6b).

In early October, the effort of raising the hangar segments up one by one began by means of a rolling scaffold which could be moved along the center line of the hangar (see image 7a). One by one, the pre-assembled hangar segments were raised and bolted together using crosspieces (see image 7b). The assembly of the hangar was completed on October 23, 1915 (see image 7c). The hangar was put up under the supervision of the K.u.K military construction department (see image 8).

Image 8 Above: Seal stamp of the K.u.K military construction department in charge of Timisoara airship hangar construction (see scharekf.hu)

At that time, Serbian pilots began conducting frequent reconnaissance flights in the area extending between the border and Timisoara. Consequently, on September 4, 1915 General Conrad asked Commander of the General Staff of the German Army, G.d.I. von Falkehayn in Pleß to secure the airship hangar construction site by providing defense cannons. The construction site was optimally secured by the military (see image 9), leading to the commencement of military operations at Szentandrás airship base in October 1915.

The airship hangar (see image 10) measured 180 meters in length, 30 meters in width, and 30 meters in height (see Navy, 1990/I, page 2).

During an additional expansion stage, the airship hangar was equipped with two large gates and the slanted walls were covered with tarpaulin.

In rapid succession, the members responsible for the construction of the airship base changed (Oberstleutnant Graf Breda/from September 1: Hauptmann Pachner von Zobor/from September 21: Hauptmann Hirsch/later: Hauptmann Lumbe Edler von Mallonitz) (see: Ploes, 2018, page 16).

Simultaneously, workshops for the maintenance and repair of airships were established along with depots for spare parts and ammunition as well as a gas work.

On September 18, 1915 command of the Mackensen army group was transferred to Timisoara for the military campaign against Serbia. The following were part of the army group:

Image 9 Left: Two Austrian-Hungarian batteries of air defense artillerists took over the construction security of Timisoara airship base (image: Racil Collection).

Image 10 Below: Hangar construction from bolted, prefabricated parts (image: Biber Archive).

The newly established deutsche 11. Armee (also known as Südarmee) under the command of Generalmajor Max von Gallwitz the K.u.K. 3. Armee under the command of General Herman Kövess von Kövessháza, and later (after Bulgaria's entry into the war) the Bulgarian 1st Army under the command of Generalleutnant Boyadzhiev.

On September 29, 1915 an order to attack was issued to the subordinated armies; the military campaign began on October 6, 1915 with the crossing of the Danube river. The city of Belgrade fell on October 9, 1915.

Bulgaria entered the war on October 14. United, the Serbian army was repulsed to Kragujevac by the end of October (see Kustan & Penz, pages 13-15).

From October 19, 1915 the German 11th Army crossed the Danube river near Belgrade. To the north, they were supported by the k.u.k. 3rd Army which crossed the Danube river near Mitrowitza (see Ploes, 2018, page 16).

Chapter 2: Royal Saxon Field Forces for Airships Nr. 14 – "LT 14" Mission Training at Schneidemühl and Dennewitz Airship Base

Image 11 Above: Stamp of Königlich Sächsischer Feldtrupp for Luftschiffe Nr. 14 (image: Biber Archive).

2.1 Military Basic Training and Aeronaut Training at Schneidemühl

As soon as the war began, a reorganization within the armed forces began which resulted in the removal of the term Luftschiffer-Bataillone, as they were modified to Luftschiffer-Ersatzabteilungen. Instead of Luftschifferkompanien, there were now Luftschiffertrupps which were assigned to each airship or airship base.

For example, 3. Kompanie (Saxon Kompanie) in Dresden was part of Luftschiff-Bataillon Nr. 2 which in turn was part of Luftschiffertruppe des Heeres. 3. Kompanie in Schneidemühl was part of Luftschiff-Bataillon Nr. 5.

Königlich Sächsischer Feldtrupp für Luftschiffe (Royal Saxon field forces for airships) Nr. 14 – "LT 14" for short, developed from 3. Kompanie (Saxon Kompanie). "LT 14" was to transfer to the new airship base in Timisoara. Their area of responsibility included:

- Gas procurement
- Airship maintenance, as well as
- Base development and upkeep.

Luftschifftrupps were not subordinate to the authority of a Luftschiffkommandant. His authority over a Luftschiffertrupp was limited to concerns relating directly to airship operations. In all other cases, the Luftschiffertrupp was under the authority of the Kommandant des Luftschiffhafens (airship base commander).

Kommandanten der Luftschiffe des Heeres in turn were subordinate to the direct authority of Oberste Heeresleitung (Supreme Army Command) or an Armeeoberkommando (Army Command).

In order to secure the operational readiness of airships and hangars alike, many aircrew and ground crew members were trained. Schneidemühl airship post was a good training facility which also served as an important starting point for war Zeppelin airship

Image 12 Above: Paul Biber cleaning his gun at the military barracks on Alaunstraße in Dresden, August 1914 (image: Biber Archive).

deployments in the east.

Sächsische Luftschiffer-Kompanie turned into Königlich Sächsischer Feldtrupp für Luftschiffe (Royal Saxon field forces for airships Nr. 14 or LT 14) (see image 11).

There is a message from Luftschiff-Trupp 14 dated August 21, 1914 regarding the successful landing of airship P IV at Dresden-Kaditz airship base (see Archival record Nr. 1033). The airship with the designation P IV is a Parseval-Schiff PL 16 which began operations in 1913. It was utilized by the Imperial German Army Administration from October 2, 1913 until March 24, 1916 (see Griehl, 2010, page 23).

In the summer of 1917, Königlich-Sächsischer Feldtrupp für Luftschiffe was utilized at a variety of airship bases in Germany, Hungary (Timisoara), and Bulgaria (Jamboli). Information taken from field post letters confirmed that "LST 14" troop allocation took place in Weiterstadt/Darmstadt district in March 1918 (see Dresden Stadtwiki).

Following the completion of basic training on August 22, 1914 (see image 12) with III. Bataillon Reserve-Infantrie-Regiment Nr. 241 (see image 13) and 1. Ersatz-Bataillon, Leib-Grenadier-Regiment Nr. 100, Paul Biber was assigned to Luftschifftrupp Nr. 14, "LT 14" on December 31, 1914 (see armed forces passport entry/Images 14–16).

Paul Biber was pleased about his transfer to Luftschifftrupp 14; he anticipated many interesting activities associated with his new role in airship support which held potential similarities to his professional tasks in civilian life. He had been interested in airships ever since his first encounter with them in Dresden, even having an opportunity to go along for a ride in both Parseval airships and particularly Zeppelin airship LZ 17, "*Sachsen*".

There is a Luftschifftrupp 14 message dated February 12, 1915, indicating that LT 14 had left for further training in Schneidemühl (see archival document Nr. 1033).

The Royal Saxon Feldtrupp für Luftschiffe Nr. 14 trained future ground crew personnel for the Szentandrás-Temesvár airship base near Schneidemühl (present day Pila in Poland/see images 17, 18, and 19).

The airship hangar was situated approximately 3 km north of the city of Schneidemühl; it was large enough to accommodate two airships (see Navy, 1992/IV, page 6).

Image 13 Above & Below: Paul Biber military passport entry (image: Biber Archive).

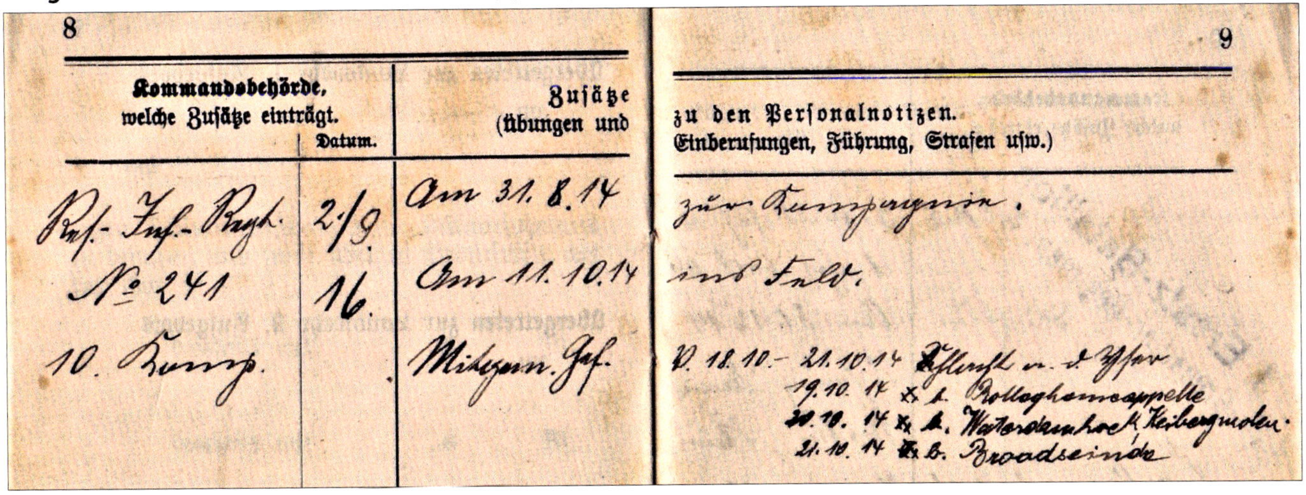

Image 14 Left: Stamp of III. Bataillon Reserve-Infantrie-Regiment Nr. 241 (image: Biber Archive).

Image 16 [Facing Page, Above]: Paul Biber military passport entry (image: Biber Archive).

2.2 Continuation of Aeronaut Training in Dennewitz

Following his training in Schneidemühl, Paul Biber left Königlich-Sächsischer Feldtrupp für Luftschiffe Nr. 14 (LT 14 for short) from late May 1915 to attend training in Dennewitz near Jüterbog.

According to a message transmitted by LT 14 on April 10, 1914, Zeppelin LZ 16 (which had been seized by the Army after successful trial operations on March 14, 1913) was transferred to Jüterbog (see archival document Nr. 0133). After many training operations, Z IV was disarmed in Jüterbog in the fall of 1916; it was considered to be outdated.

There is an additional message from Luftschiff-Trupp 14, dated May 11, 1915; it indicated that LT 14 was transferred to Jüterbog for further training operations (see archival document Nr. 1033).

During his stays both at Schneidemühl and Dennewitz, Paul Biber was trained and prepared in a variety of activities typically carried out by ground crew members in airship operations although he appeared to have expected more from his training.

From June 1915 Paul Biber was accommodated at the home of Mr. and Mrs. Höhne (siehe Abb. 20). The Höhne family lived in a small house in Dennewitz on Wittenberger Straße 7 (see image 21). The couple had seven children (see image 22).

As part of an unfortunately incomplete text (cut off on the right-hand side) from his field postcard (sent from Dennewitz, dated August 8, 1915) in image 19, Paul Biber congratulated his parents and younger siblings on their move into a new home. "Hopefully you will soon settle into the new nest and you will have a lovely time there."

The right-hand side (see underlined text) of the field postcard lists some interesting facts regarding Paul Biber's service with LT 14. Here, Paul Biber mentioned that he "had an opportunity to work on the airship". From the remainder of the text, one can possibly conclude that he may have

Image 17 [Facing Page, Below]: Schneidemühl field postcard, dated April 12, 1915/ Paul Biber underlined the towns and locations he had already visited (image: Biber Archive).

Image 15 Left: Paul Biber's military passport (image: Biber Archive).

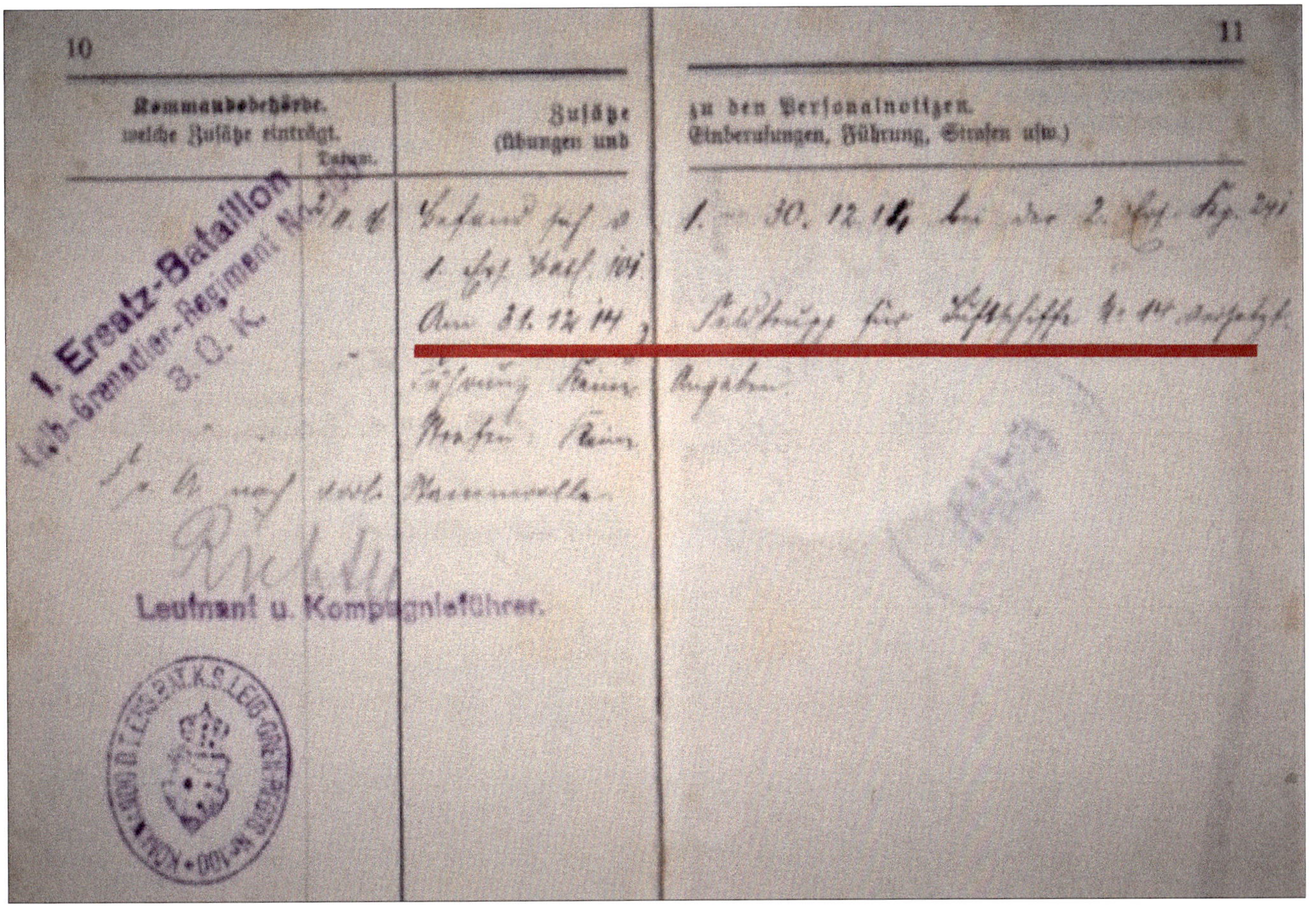
10
11
Kommandobehörde, welche Zusätze einträgt.
Datum.
Zusätze (Übungen und
zu den Personalnotizen. Einberufungen, Führung, Strafen usw.)
1. Ersatz-Bataillon
Leib-Grenadier-Regiment
8. O. K.
Leutnant u. Kompagnieführer.

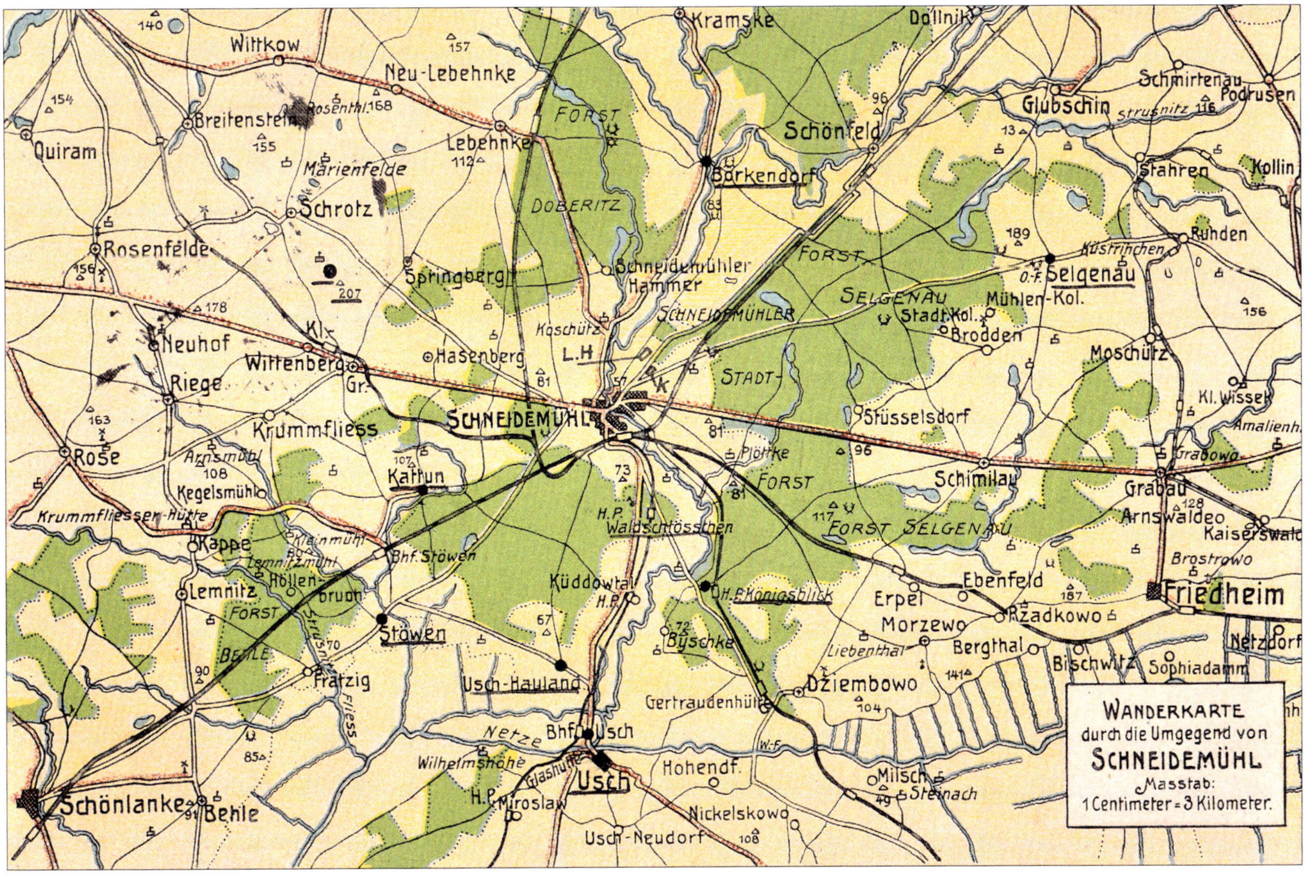
WANDERKARTE
durch die Umgegend von
SCHNEIDEMÜHL
Masstab:
1 Centimeter = 3 Kilometer.
Schneidemühl
Wittkow
Neu-Lebehnke
Kramske
Dollnik
Quiram
Breitenstein
Lebehnke
Schönfeld
Glubschin
Schmirtenau
Podrusen
Marienfelde
Borkendorf
Stahren
Kollin
Schrotz
Doberitz
Ruhden
Rosenfelde
Springberg
Schneidemühler Hammer
Selgenau
Mühlen-Kol.
Neuhof
Stadt-Kol.
Brodden
Riege
Wittenberg
Hasenberg
Moschütz
Kl. Wisser
Krummfliess
Stüsselsdorf
Amalienhof
Rose
Kegelsmühl
Kattun
Plöttke
Schmilau
Grabau
Krummfliesser-Hütte
Arnswaldeo
Kappe
Bhf. Stöwen
Kaiserswalde
Lemnitz
Küddowtal
Ebenfeld
Brostrowo
Stöwen
Erpel
Zadkowo
Friedheim
Morzewo
Netzdorf
Byschke
Bergthal
Bischwitz
Sophiadamm
Pratzig
Usch-Hauland
Dziembowo
Gertraudenhütte
Netze
Bhf. Usch
Wilhelmshöhe
Usch
Hohendf.
Milsch
Steinach
Schönlanke
Behle
Miroslaw
Nickelskowo
Usch-Neudorf
FORST
FORST SELGENAU

Königsblick b. Schneidemühl.

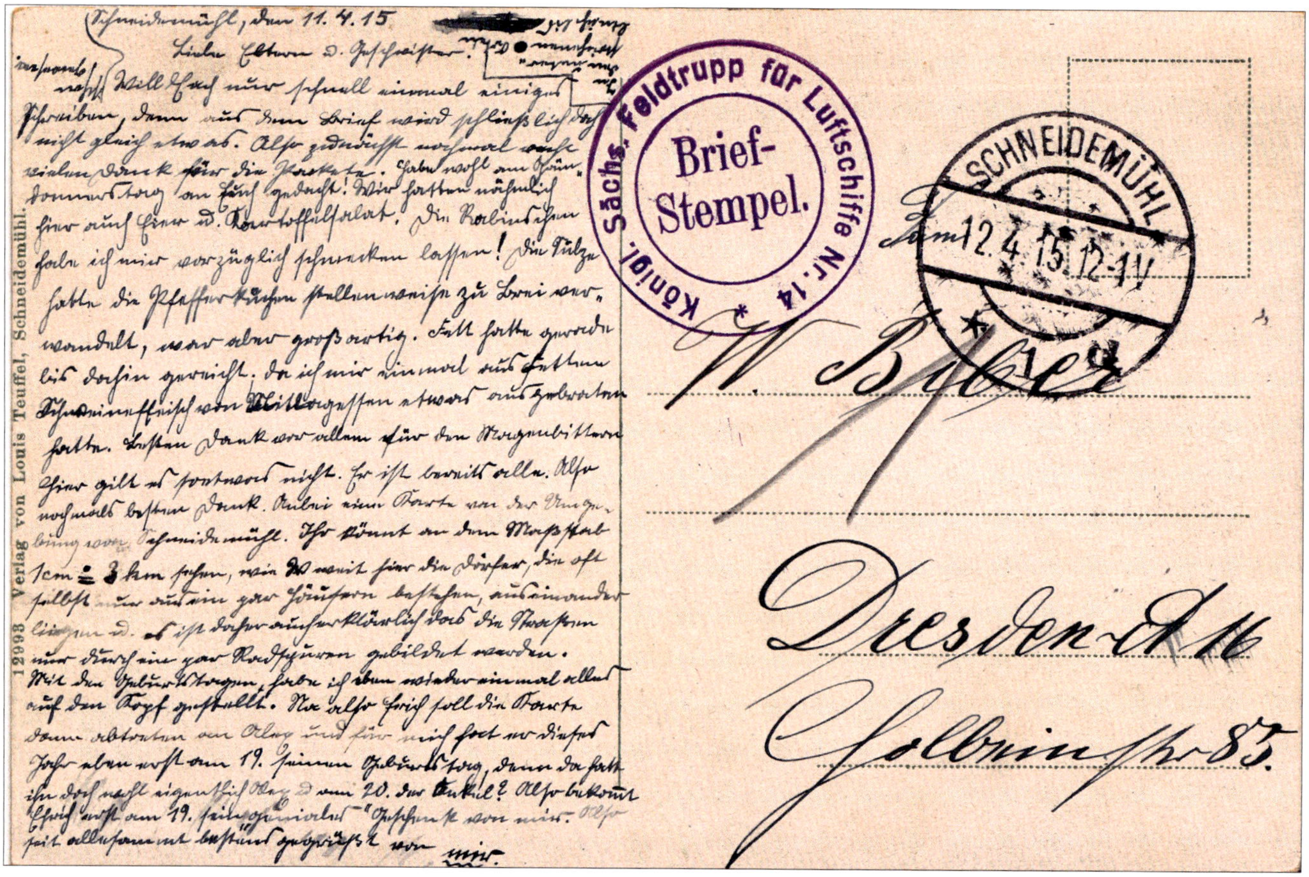
12993 Verlag von Louis Teuffel, Schneidemühl.
Schneidemühl, den 11.4.15.
Königl. Sächs. Feldtrupp für Luftschiffe Nr. 14
Brief-Stempel.
SCHNEIDEMÜHL
12.4.15. 12-1V
Dresden A 16

Image 20 Above: Frau and Herr Höhne – Paul Biber's hosts during his stay in Dennewitz (image: Biber Archive).

Image 18 Facing Page, Above: Königsblick postcard near Schneidemühl (image: Biber Archive).

Image 21 Right: Reverse of image 16. "My hosts in Dennewitz. Herr and Frau Höhne, Wittenberger Straße, June 1915." (image: Biber Archive).

Image 19 Facing Page, Below: Field postcard (reverse/image 8) written by Paul Biber, dated April 12, 1915 (image: Biber Archive).

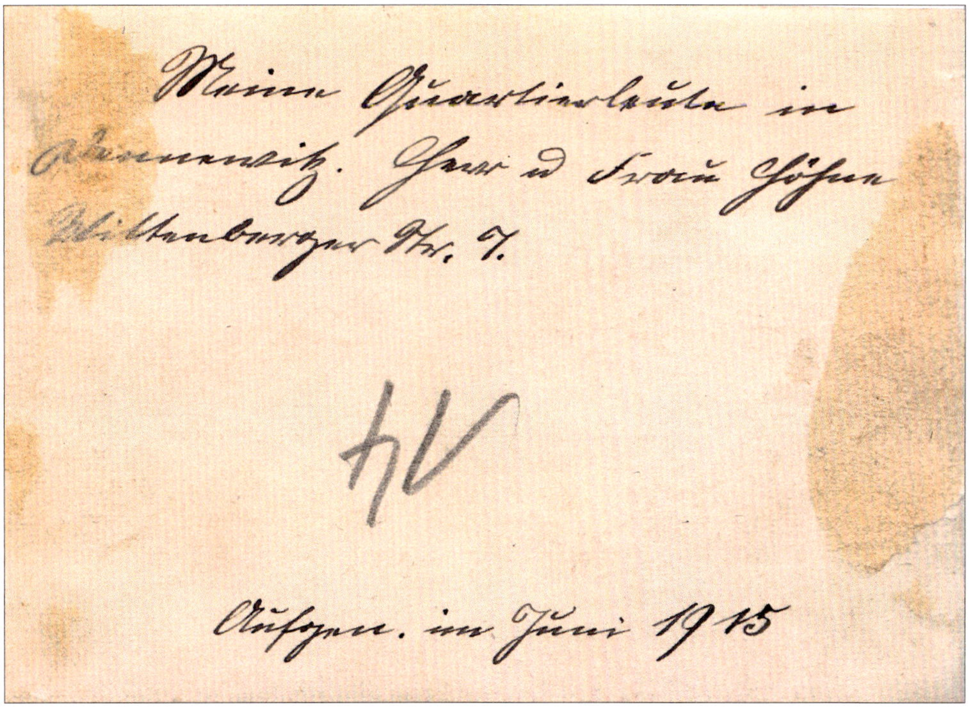

Meine Quartierleute in
Dennewitz. Herr u. Frau Höhne
Wittenberger Nr. 7.

Aufgen. im Juni 1915

Image 22 Above: Frau Höhne and her seven children (image: Biber Archive).

participated in at least one trip onboard an airship... "unsuccessfully searched for a landing spot...", "...happily arrived. (I) have already completed.... rides...", "...(will ride) again tomorrow at 0330 am (wake up at 0100)".

Additional photographs from his army life in Dennewitz can be found in **images 24-40.** Image 25 shows the Dennewitz airship hangar in the background. Building a sand wall surrounding the bomb shed provided "LT 14" soldiers with an additional occupation for several weeks. On the reverse of the image (see image 28), a note reads: on Saturday, August 13 while chopping wood in the field kitchens at the Dennewitz school courtyard. I spent several days there, diligently chopping wood for the field kitchens with my old, faithful comrades Hoffmann (spelled with two f's – ff), Hauch, Sauer, and Groß II. The school's icehouse is pictured in the background.

Unfortunately, we had to bury Bob, our loyal Kompanie dog on June 4, 1915. A car brought an abrupt ending to his life (see images 29 and 30).

Staffing airship operations were geared toward assigning one Landungstrupp (landing troop/ landing forces) and one Schiffspflegegruppe (airship maintenance group) to each airship base. The troops consisted of personnel well suited for the execution of upcoming tasks; troop members were specifically chosen because they were thought to meet these criteria. Training of personnel was activated at the outbreak of the war and jointly with the serial production of military airships and the expansion of the airship hangar and base network.

At its peak, utilization of airship bases during the war counted approximately 40 Luftschifftrupps. The average size of a troop amounted to approximately 140 people (see Navy, 1992/I, page 10).

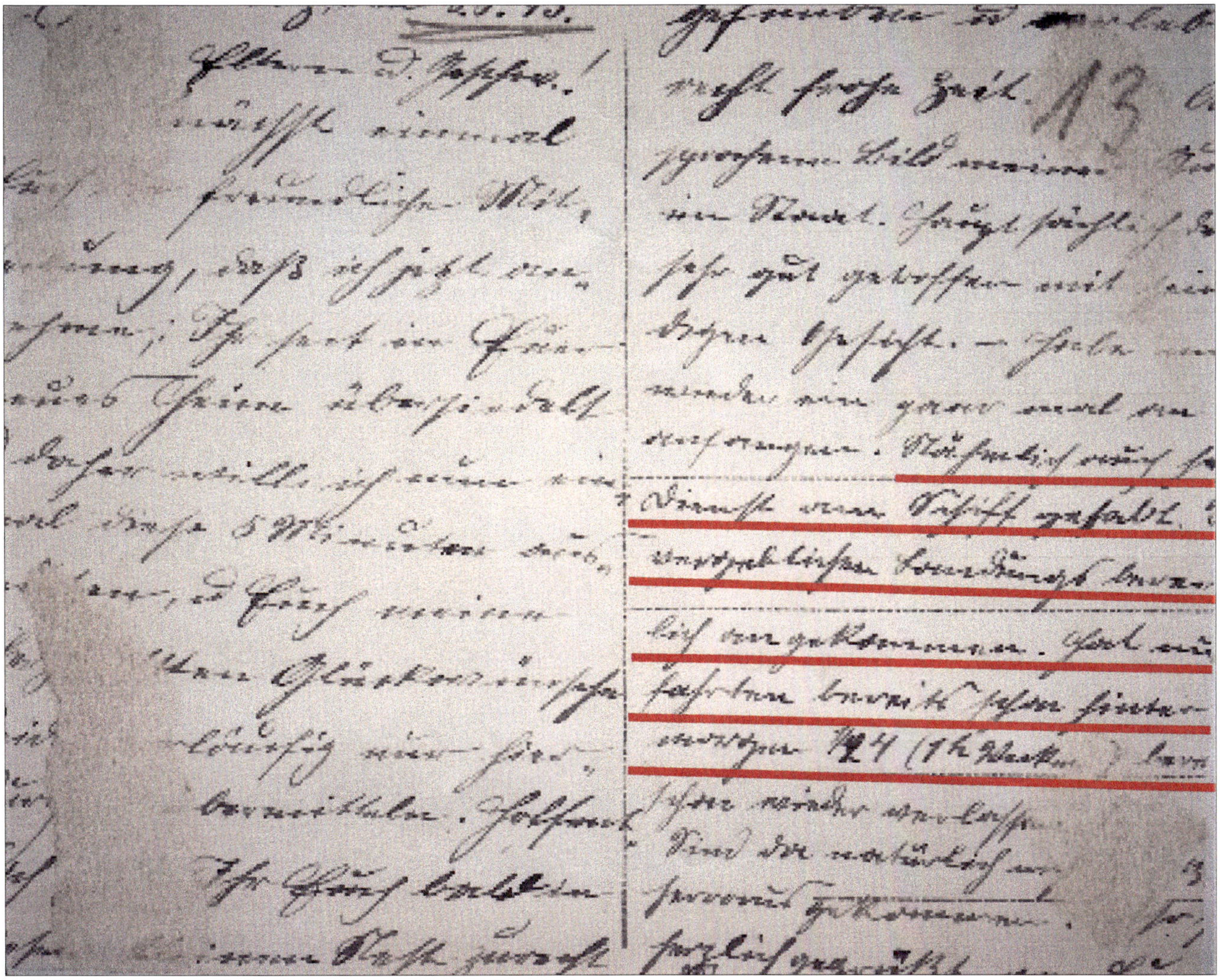

Image 23 Above: Reverse side of photograph pictured in image 22 containing a report from Dennewitz dated August 8, 1915 (image: Biber Archive).

Image 24 Right: Paul Biber (3rd from left) loading gas cylinders in Dennewitz on June 11, 1915 (image: Biber Archive).

Image 27 Above: Chopping wood at the field kitchen in the courtyard of Dennewitz training facility (image: Biber Archive).

Image 25 Facing Page, Above: Paul Biber standing guard at the bomb shed on July 2, 1915; the Dennewitz airship hangar is pictured in the background. Building the surrounding sand wall provided "LT 14" soldiers with an additional occupation for several weeks (image: Biber Archive).

Image 26 Facing Page, Below: Soldiers prior to reporting for drill in front of the training building in 1915 (image: Biber Archive).

Image 28 Below: Reverse of photograph pictured in image 27 (image: Biber Archive).

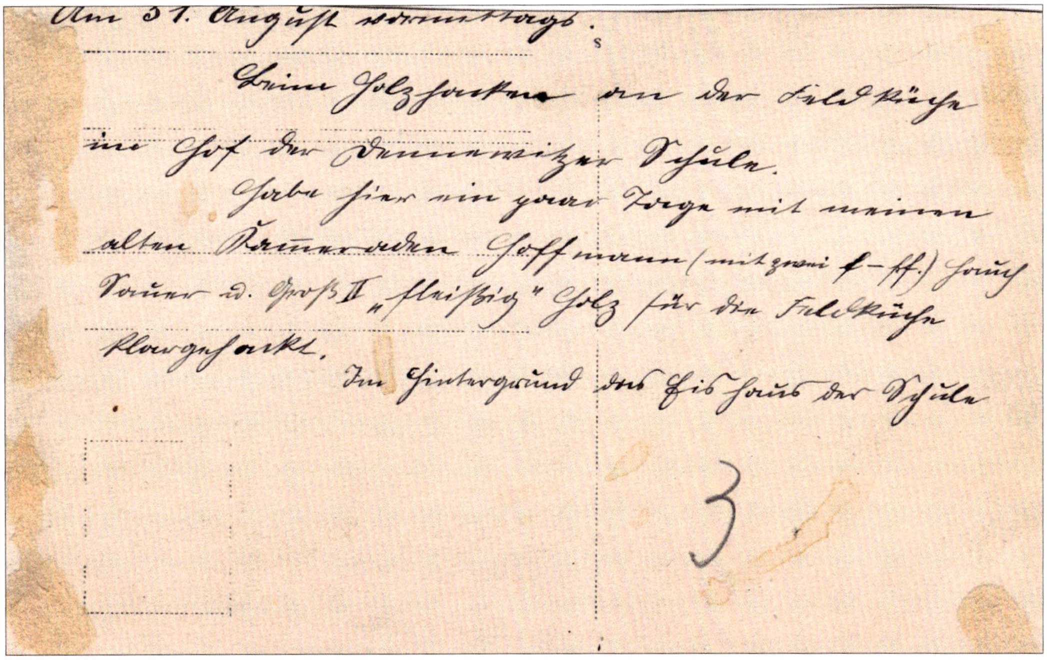

Am 31. August vormittags.

Beim Holzhacken an der Feldküche im Hof der Dennewitzer Schule.

Habe hier ein paar Tage mit meinem alten Kameraden Hoffmann (mit zwei f-ff.) [illegible] Sauer d. Gross II „fleißig" Holz für die Feldküche kleingehackt.

Im Hintergrund das [illegible] der Schule

3

Image 29 Above: June 4, 1915 – Bob's grave. He had been our company dog since Schneidemühl. He was hit by a car (see text in image 28/image: Biber Archive).

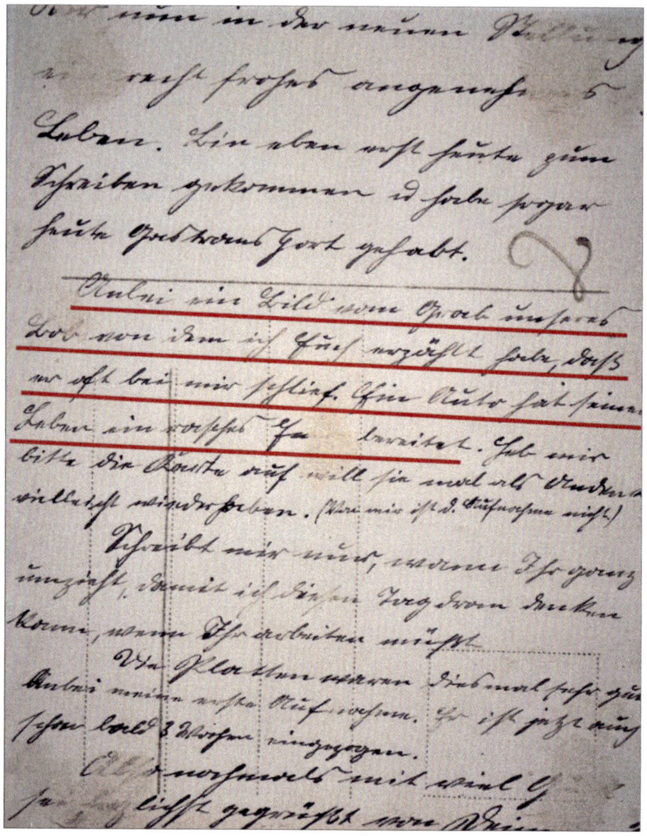

Image 30 Left: Reverse of photograph pictured in image 29 (image: Biber Archive).

Ground personnel of Königlich Sächsische Feldtrupps für Luftschiffe LT 14 were divided into two groups, which were trained for two key activities. Each troop was trained to provide focused support of the other troop in case of an emergency.

"Landungstrupp" (Haltemannschaft)

(Landing Crew)

Tasks:

1. Preparation and execution of "Luftschiff starten" phase (Airship startup/departure)

- "Luftschiff aushallen!" (move airship out of the hangar)
- "Luftschiff in Position bringen!" (move airship into position)
- "Luftschiff hoch!" (airship "up/departure")

2. Preparation and execution of "Luftschiff landen" phase (Airship landing)

- "Luftschiff – Halteseile erfassen!" (airship – grab retaining ropes)
- "Luftschiff zur Halle – marsch!" (move airship to the hangar)
- "Luftschiff einhallen!" (move airship into the hangar)

"Schiffspflegetrupp"

(Ship Maintenance Team)

Tasks:

1. Airship care, maintenance, and repairs
2. Loading and equipping of airships
3. Airship hangar care, maintenance, and repairs, including the functionality of the large doors
4. Execution of tasks for maintaining airship operations (i.e., power generation for gas production, lighting of airship hangars, workshops, and barracks, as well as navigational beacons and Funkentelegraphie)
5. Instrument care, maintenance, and repairs (i.e., clocks, pressure gauges, anemometers, and voltmeters).

Image 31 Above: View of Dennewitz including airship hangar; mid-August 1915 (image: Biber Archive).

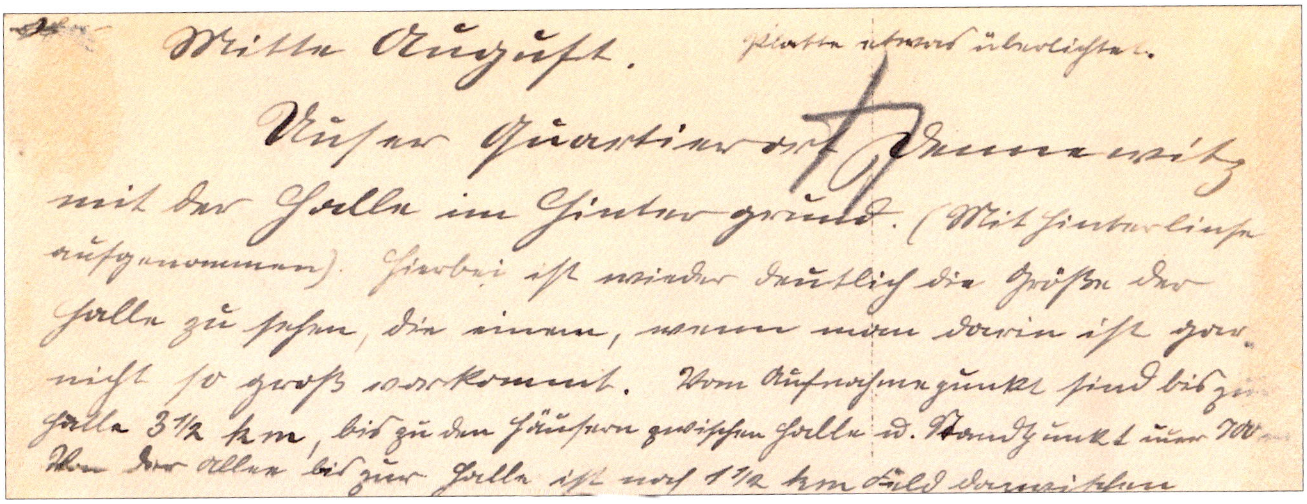

Mitte August. Platte etwas überbelichtet.

Unser Quartierort Dennewitz mit der Halle im Hintergrund. (Mit Hinterlinse aufgenommen). Hierbei ist wieder deutlich die Größe der Halle zu sehen, die einem, wenn man darin ist, gar nicht so groß vorkommt. Vom Aufnahmepunkt sind bis zur Halle 3 1/2 km, bis zu den Häusern zwischen Halle u. Standpunkt nur 700 m. Von der Allee bis zur Halle ist noch 1 1/2 km Feld dazwischen

Image 32 Above: Reverse of photograph pictured in image 31 (image: Biber Archive).

Mitte August

Unser Quartierort Dennewitz mit der Luftschiffhalle im Hintergrund (mit Hinterlinse aufgenommen). Platte etwas überbelichtet. Hierbei ist wieder deutlich die Größe der Halle zu sehen die einem, wenn man darin ist, gar nicht so groß vorkommt. Vom Aufnahmepunkt sind bis zur Halle 3 ½ km, bis zu den Häusern zwischen Halle und Standpunkt nur 700 m. Von der Allee bis zur Halle ist noch 1 ½ km Feld dazwischen.

Image 33 Above: Text written on reverse (see image 32/image: Biber Archive).

Image 34 Facing Page Above: Converting and reconnecting gas cylinders at the gas shed near Niedergörsdorf, July 18, 1915 (image: Biber Archive).

Image 35 Facing Page Below: Moving gas cylinders from the wagon into a vehicle at the railway station (image: Biber Archive).

Image 36 Above: Paul Biber (left), carrying out "chamber work" by shining his boots (using gun grease) (image: Biber Archive).

Image 37 Right: Comrade Schwenkler. (image: Biber Archive).

Image 38 Above: Comrade Delli. (image: Biber Archive).

Image 39 Above: Comrade Eismann in front of the truck. (image: Archiv Biber).

Image 41A Above: Paul Biber carries out anemometer repairs on the roof of an airship hangar near Niedergörsdorf. (image: Biber Archive).

Image 41B Above: Paul Biber carries out anemometer repairs on the roof of an airship hangar near Niedergörsdorf. The avenue, the new Jüterbog warehouse, and the gasworks are pictured in the background; in front of it was the old landing area (image: Biber Archive).

Whenever an opportunity presented itself, Paul Biber made use of his skills as a precision engineer. He was responsible for maintaining the anemometer on top of the airship hangars. Images 41 and 42 show impressions from the daily life of Paul Biber's military service.

Image 40 Above: Comrade Deckelmann. (image: Archiv Biber).

Image 42A Above: A functioning anemometer following Paul Biber's successful repairs (image: Biber Archive).

Image 42B Above: A functioning anemometer following Paul Biber's successful repairs (image: Biber Archive).

3. October 1915 – "LT 14" Deployment to Timisoara Airship Base

Image 43 Above: Luftschiffer Paul Biber, member of Königl. Sächs. Feldtrupp für Luftschiffe Nr. 14 – "LT 14" ready for deployment (image: Biber Archive).

Image 44 Above: Stamp of "KÖNIGL. SÄCHS. FELDTRUPP FÜR LUFTSCHIFFE NR. 14" (image: Biber Archive).

3.1 Train Ride to Timisoara Airship Base with a Stopover in Kistelek

In the campaign against Serbia, Germany supported overstrained allies Austria-Hungary and Turkey.

"After the Entente breached Greece's neutrality on October 5, 1915 by offloading troops in Saloniki in order to support the Serbian front" (Mückler, 2013, page 106), the Central powers were forced to react.

As attacks on Saloniki were only possible from the air, possible attack opportunities had to be verified using airships or aircraft.

In this way, Paul Biber was issued orders to transfer to his new duty station, "Szentandrás-Temeswar" airship base. Aeronaut Paul Biber, member of Königlich Sächsischer Feldtrupp für Luftschiffe Nr. 14 – "LT 14", ready for deployment (see image 43). Paul Biber and other "LT 14" military troops (see stamp in image 44) were loaded onto railway cars along with all military equipment in order to promptly be transported to southern Hungary.

A message transmitted by Luftschiff-Trupp 14 on October 10, 1915 reads that LT 14 was relocating to Timisoara by order of the chief of aeronautics (see archival document Nr. 1033).

Every now and then the train stopped at a railway station, for instance in Brünn, where the troops could purchase cigarettes, drinks, or request other types of assistance if needed **(see image 45).**

Image 45 Above: "LT 14" train stop at Brünn station (image: Biber Archive).

One of the more memorable stops that Luftfeldtrupp Nr. 14 experienced on their journey to Szentandrás-Temesvár was when their train stopped at the Kistelek station on October 21, 1915. The town of Szeged was approximately 20 km away from the railway station.

Upon arrival at Kistelek, the search began for the correct position to capture photographs (see image 46) of the stopped train (see image 47).

As it appears, searching for just the right spot turned into a fairly good motif for a photograph as well. The note on the reverse of the stopped train photograph reads: our train at the Kistelek station (see image 48).

The large camera as well as the search for the proper motif and position were a reflection of the level of professionalism upheld while capturing and developing images.

The news of Königlich-Sächsischer Feldtrupp für Luftschiffe Nr. 14's arrival spread quickly throughout the city. People came in flocks to welcome the German soldiers (see image 49). Paul Biber's notes on the subject (see image 50) indicate that people brought all available musical instruments in order to "make noise".

The photograph (see image 51) shows two wide sections of the road. The note on the reverse of the photograph (see image 52) indicates that this type of road construction was typical of the area which was a generally slightly aqueous environment.

The first part of the road extends to the water ditch, where the other part of the road begins. The footpaths on either side of the road were paved with bricks in order to provide a better walking path during the rain so as to avoid getting stuck in the mud.

As the stop at Kistelek was supposed to last several hours, the soldiers were invited to people's farms. The officers and soldiers were quickly supplied with food and drinks (see images 53-56).

The Luftschiffer took a close look at the farms (see image 52). Paul Biber particularly liked the clean way in which the walls beneath the alleyway were painted on most houses (see note on **image 56).**

The locomotive whistled multiple times as everyone returned to the military train. Soon after, the train continued its journey toward Szeged and on to Szentandrás-Temesvár.

Image 46 Above: Proper position for image capture as viewed by Paul Biber (2nd from right/ image: Biber Archive).

Image 47 Below: Arrival of the "LT 14" train at Kistelek station on October 21, 1915 (image: Biber Archive).

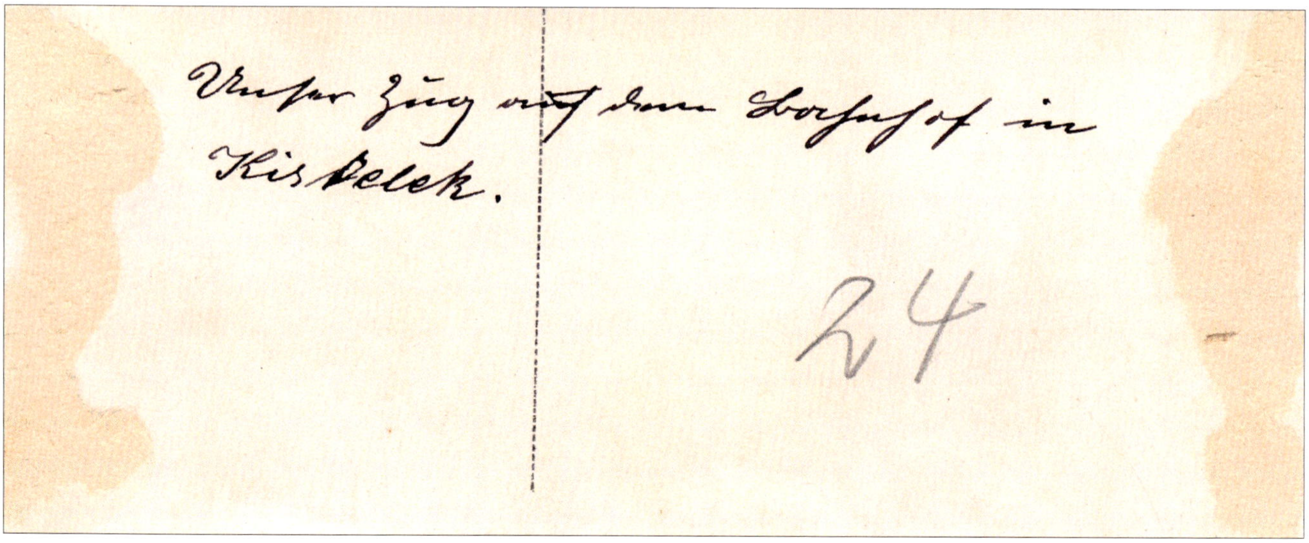

Image 48 Above: Reverse of image 47 (Biber Archive).

Image 49 Below: The Kistelek population welcomes "LT 14" in front of the tavern (image: Biber Archive).

Image 50 Facing Page, Above: Reverse side of the above photograph (image 49/ Biber Archive).

Image 51 Facing Page, Below: Kistelek church with a road characteristic of the area (image: Biber Archive).

21. 10. 15.

Einige Kameraden vor einem Gasthof in Kistelek. Sämtliche verfügbaren Musikinstrumente wurden herausgeräumt u. damit Krach gemacht.

(Es gab hier in Kistelek einen sehr guten Wein! Habe solchen hier noch nicht wieder zu trinken bekommen.)

21

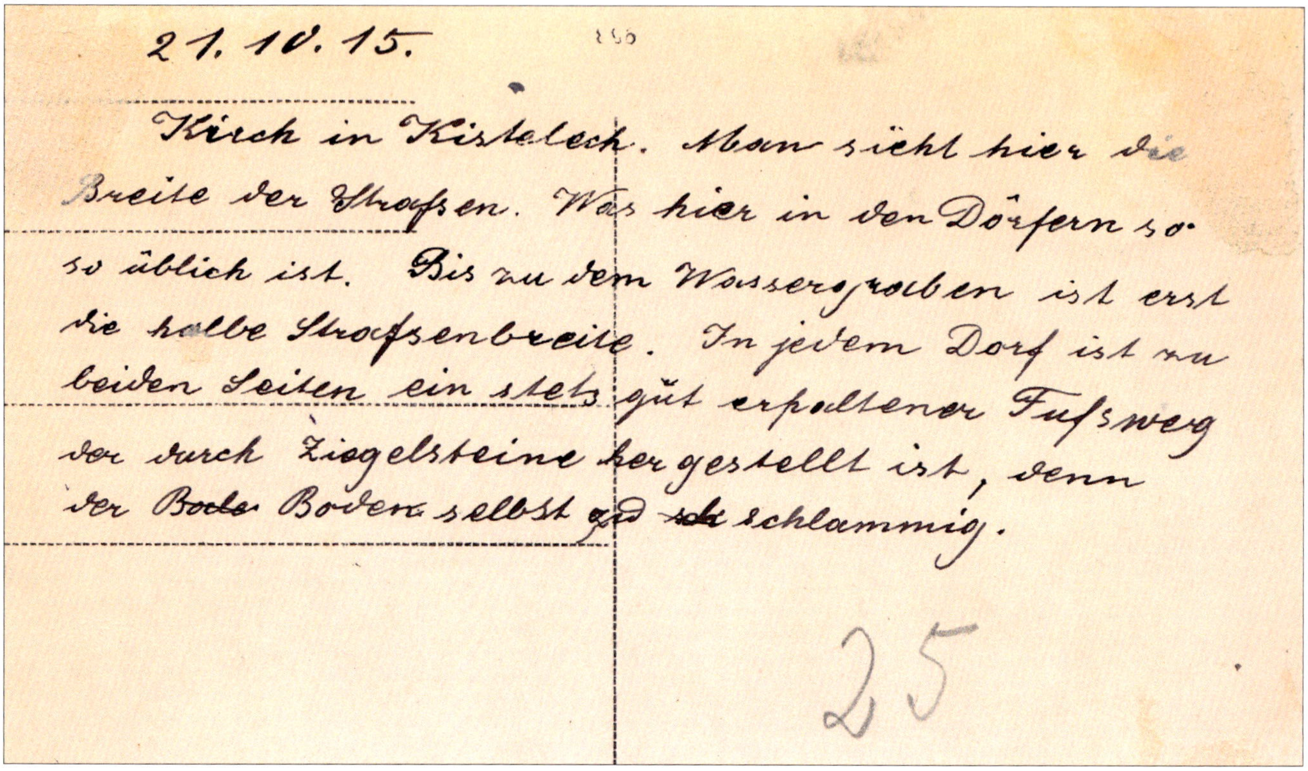

21. 10. 15.

Kirch in Kistelech. Man sieht hier die Breite der Straßen. Was hier in den Dörfern so so üblich ist. Bis zu dem Wassergraben ist erst die halbe Straßenbreite. In jedem Dorf ist zu beiden Seiten ein stets gut erhaltener Fußweg der durch Ziegelsteine hergestellt ist, denn der ~~Bode~~ Boden selbst ~~ist~~ ~~ist~~ schlammig.

25

Image 52 Above: Reverse side of the above photograph (image 51/Biber Archive).

Image 53 Below: Mother Esztike and her daughter, Kasza in front of their home (image: Biber Archive).

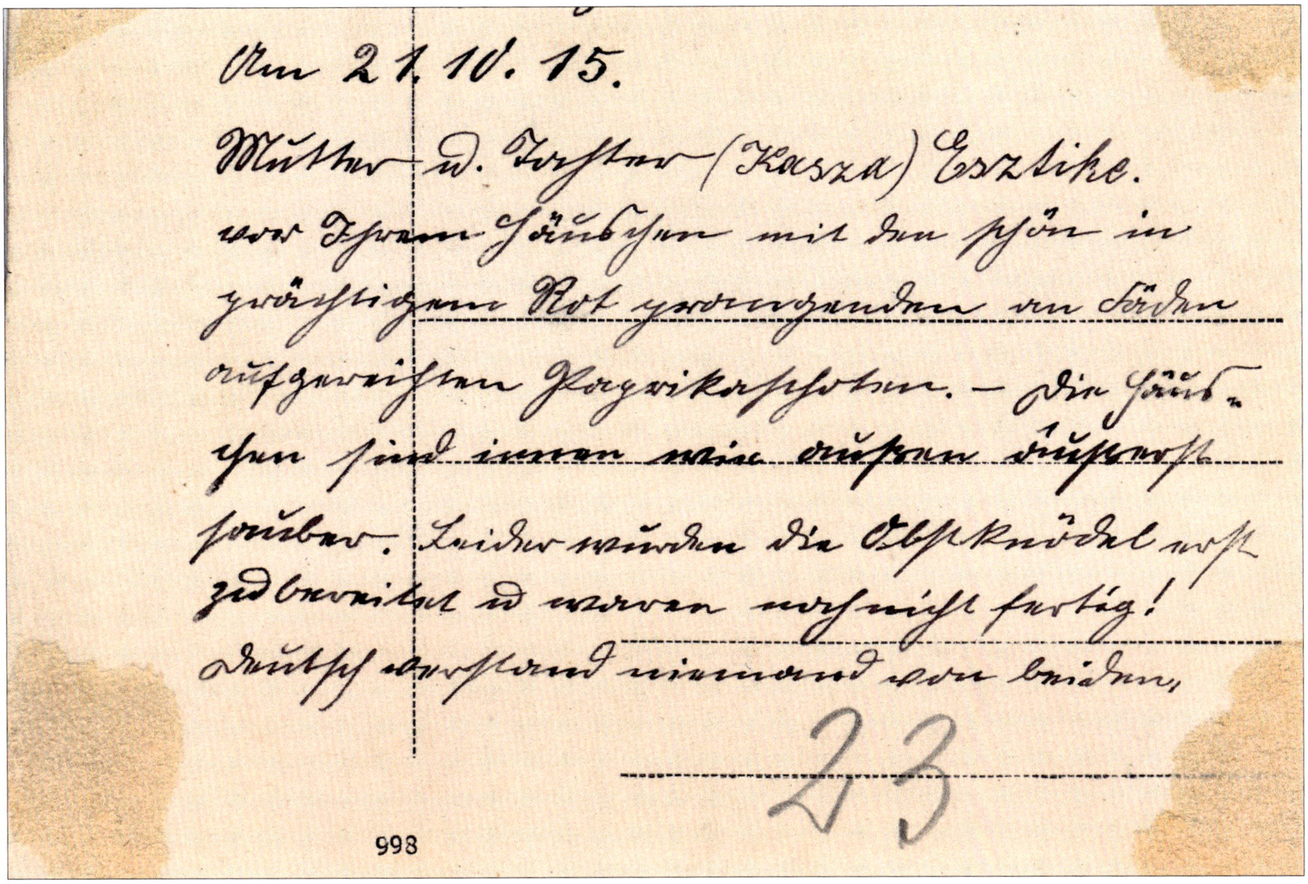
Am 21.10.15.
Mutter u. Tochter (Kasza) Esztike.
vor ihrem Häuschen mit den schön in
prächtigem Rot prangenden an Fäden
aufgereihten Paprikaschoten. – Die Häus-
chen sind innen wie außen äußerst
sauber. Leider wurden die Obstknödel erst
zubereitet u. waren noch nicht fertig!
Deutsch verstand niemand von beiden,
23
998

Image 54 Above: Reverse of the above photograph (image 53/ Biber Archive).

Image 55 Below: Arrival Text from image 54 (Biber Archive).

„Vom Aufenthalt in Kistelek am 21.10.15

Mutter und Tochter (Kasza) Esztike vor ihrem Häuschen mit den schön in prächtigem Rot prangenden, an Fäden aufgereihten Paprikaschoten. Die Häuschen sind innen wie außen äußerst sauber. Leider wurden die Obstknödel erst zubereitet u. waren noch nicht fertig! Deutsch verstand niemand von den beiden.“

3.2 October 1915 – "LT 14" Aids in the Construction of an Airship Hangar

The establishment of Timisoara airship base, bearing the code name "Adebar" required the delivery of a vast number of different supplies. Friedrich Seifert, the Sanktandres railway station chief at the time, proposed the construction of a separate 1253 m long railway to the airship hangar construction site. The proposal was accepted. "The construction of the new railway was finished in three days; on September 9, 1915 the first materials train car transported the equipment necessary to drill four artesian wells to the construction site. A total of 316 railway cars filled with construction materials were transported to the construction site in September 1915; in October, the number of cars delivering materials for the construction of the airship hangar totaled 560..." (SANKTANDRES/1981, page 164).

The new railway was also utilized to transport LT 14 officers and soldiers, materials for the equipment of workshops, as well as transport vehicles. As some buildings were ready for occupancy, it was possible to distribute some of the offloaded goods among the respective buildings.

Offiziere, Unteroffiziere, and Soldaten quickly

Image 54 Above: A typical farm of this region with a note on the reverse: "The walls beneath the covered alleyway of the house have been carefully painted with ornamentations (clearly visible near the left door)" (image: Biber Archive).

tried to put the airship base into operation in preparation for the arrival of Zeppelin airship LZ 81.

Hauptmann Hans Hirsch was in command of "ADEBAR" airship base until October 27, 1915; thereafter, Hauptmann Karl Lumbe Edler von Mallonitz took over command of "ADEBAR" (see Kustan & Penz, page 15).

While Szentandrás airship base was maintained by a k.u.k. Kommando, regular operations were not possible utilizing their personnel. German aeronauts with appropriate training were responsible for the operation of airships. Königlich Sächsischer Feldtrupp für Luftschiffe Nr. 14 – "LT 14" arrived at Szentandrás in good time **(see image 57)** to aid in securing the completion of hangar construction and to commence operations at the airship base.

By the time Königlich Sächsischer Feldtrupp für Luftschiffe Nr. 14 – "LT 14" arrived in Szentandrás, all hangar segments had been erected and interconnected using multiple longitudinal bracings (see image 58).

The next task consisted of protecting the interior by building a cover made of tarpaulin and installing two large doors in the front of the building. The rolled-up tarpaulin required to cover the hangar was made of waterproofed linen cloth. It was conveniently delivered via railway (see image 59).

Many hands were required in order to transport the heavy, rolled-up tarpaulin to the unfinished airship hangar and to attach them to the large hangar structure in a stable manner (see image 60a). Most of the sailcloth rollers were stored in barracks to protect them from the rain. Soon, progress was visible as the first sheets of tarpaulin were stretched between individual hangar arc segments (see images 61, 62a, and 62b).

There was a lot of work to be completed on the interior of the hangar as well. Tasks included anchoring the left and right tracks into the hangar floor, installing interior lighting, and reinforcing the hangar floor (see images 63 and 65).

In spite of being housed in a hangar, the airship was anchored to the hangar using ropes. In order to safely move the airship into and out of the hangar, two sets of tracks were installed in the hangar floor in longitudinal direction (one set of tracks on the left side and one on the right side) (see images 63 and 65). The airship was attached to these tracks using ropes in order to secure it until its next trip.

The tracks leading to the outside premises are

Image 57 Above: Arrival of Königlich Sächsischer Feldtrupp für Luftschiffe in Timisoara (image: Biber Archive)

Image 58 Above: Hangar framework at the time of LT 14's arrival in Timisoara in October 1915 – hangar construction in progress (image: Biber Archive).

Image 59 Above: Off-loading rolled-up tarpaulin sheets from railway cars (image: Racil Collection).

Image 60a- Above: Tarpaulin sheets are inserted into the hangar framework (see scharekf.hu).

Image 60b Above: Rolled-up tarpaulin was delivered via railway cars and temporarily stored inside barracks; it would soon be stretched between the arcs of the hangar framework. Mid-October (image: Fabry Archive).

Hangar framework at the time of LT 14's arrival in Timisoara about 25 October 1915 hangar construction in progress (image: Biber Archive).

Image 62a Above: Visible progress while stretching tarpaulin sheets between arcs of the hangar framework, Szentandrás airship hangar, late October 1915 (image: Biber Archive).

Half of Timisoara airship hangar covered by stretched tarpaulin, October 25, 1915 (image: Fabry Archive).

Image 63 Above: Working on the interior of Timisoara airship hangar, late October 1915 (image: Biber Archive).

Image 64 Below: The left tracks are easily discernable on the hangar floor (see scharekf.hu).

Image 65 Facing Page, Above: Reverse of photograph in image 63, captioned "This photograph was not taken by me. Hangar construction in progress". (Biber Archive)

Image 66 Facing Page, Below: View from the upper hangar aisle; the treads of the tracks leading onto the premises are clearly visible (Schiene/image: Biber Archive).

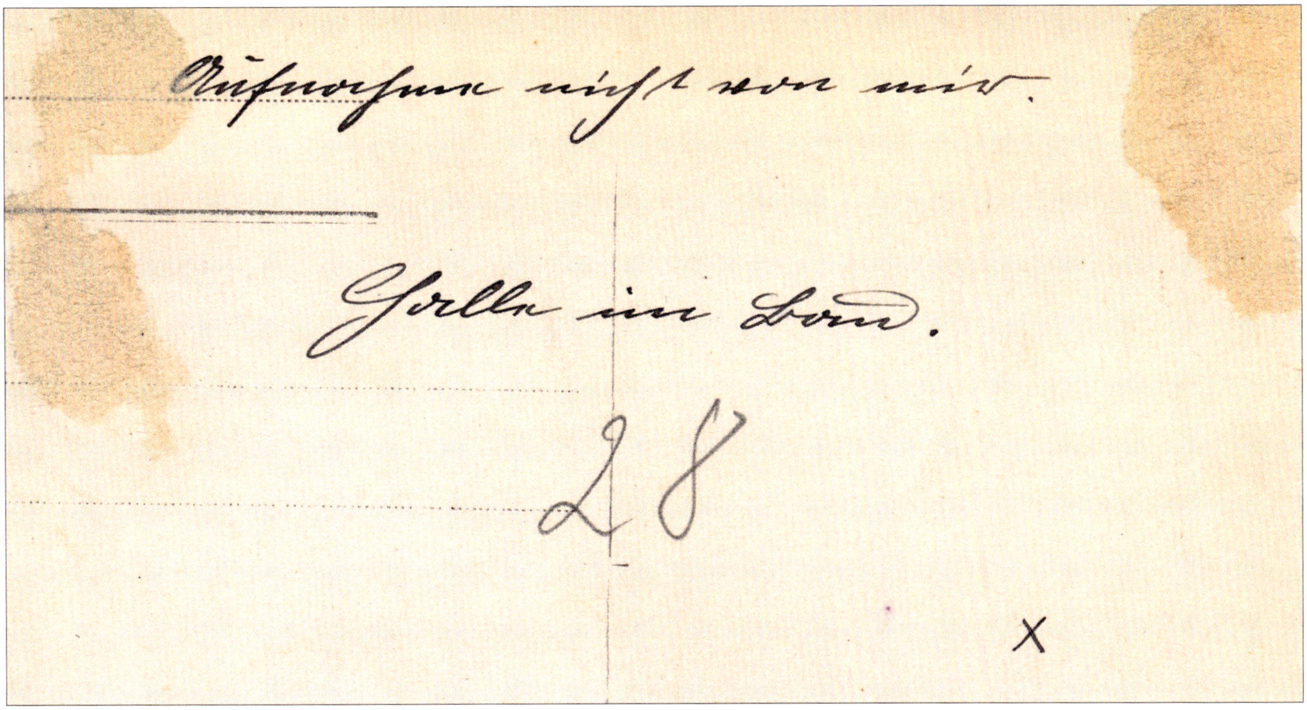
Aufnahme nicht von mir.

Halle im Bau.

28

x

Image 67a Above: The large hangar doors were attached to Timisoara airship hangar on October 28, 1915.

Image 67b Above: Timisoara airship hangar with both pairs of sliding doors (image: Biber Archive).

Image 68 Left: Door system functionality test (image: Biber Archive).

clearly visible from the upper hangar aisle (see image 66). These tracks were utilized to pull the airship by means of ropes in order to quickly anchor it outside, especially in gusty conditions.

In an effort to protect the airships from the wind and other environmental elements, the hangar was fitted with two large sliding doors which were attached to the buildings' transverse sides that had previously been left open. The functionality of the doors had been certified prior to installation (see images 67a, 67b, and 68).

3.3 November and December 1915 – Setup of Workshops, Warehouses, and Accommodations

Construction work was progressing fairly rapidly, allowing all workshops, warehouses, and accommodation barracks to be set up in accordance with their respective operational functions. On October 28, 1915 all crucial work of the Szentandrás-Temeswar airship hangar construction project was complete. The naval ensign was proudly hoisted onto the roof of the hangar **(see image 69).**

Due to the significant gas consumption required by the airships' war patrol flights to various war zones, there were periodic supply shortages of required gas cylinders. As a result, emphasis was placed on hydrogen production when constructing Timisoara airship hangar. This in turn required the autonomous generation of direct current. Electrical energy was required for other tasks on the airship base, including the operation of radio equipment as well as operation of spotlights. It was apparent that several other buildings needed to be built in addition

Image 69 Right: Hoisting the naval ensign on the roof of the Timisoara airship hangar (image: Biber Archive).

Image 70 Below: Aerial view of Timisoara airship base (see Navy, 1990/I, page 8) with identification markings.

Image 71 Above: View from the airship hangar of various buildings including workshops, warehouses, and accommodations which were built directly alongside the airship hangar at Timisoara airship base (image: Biber Archive).

Image 72 Below: Reverse side of image 71 (Biber Archive).

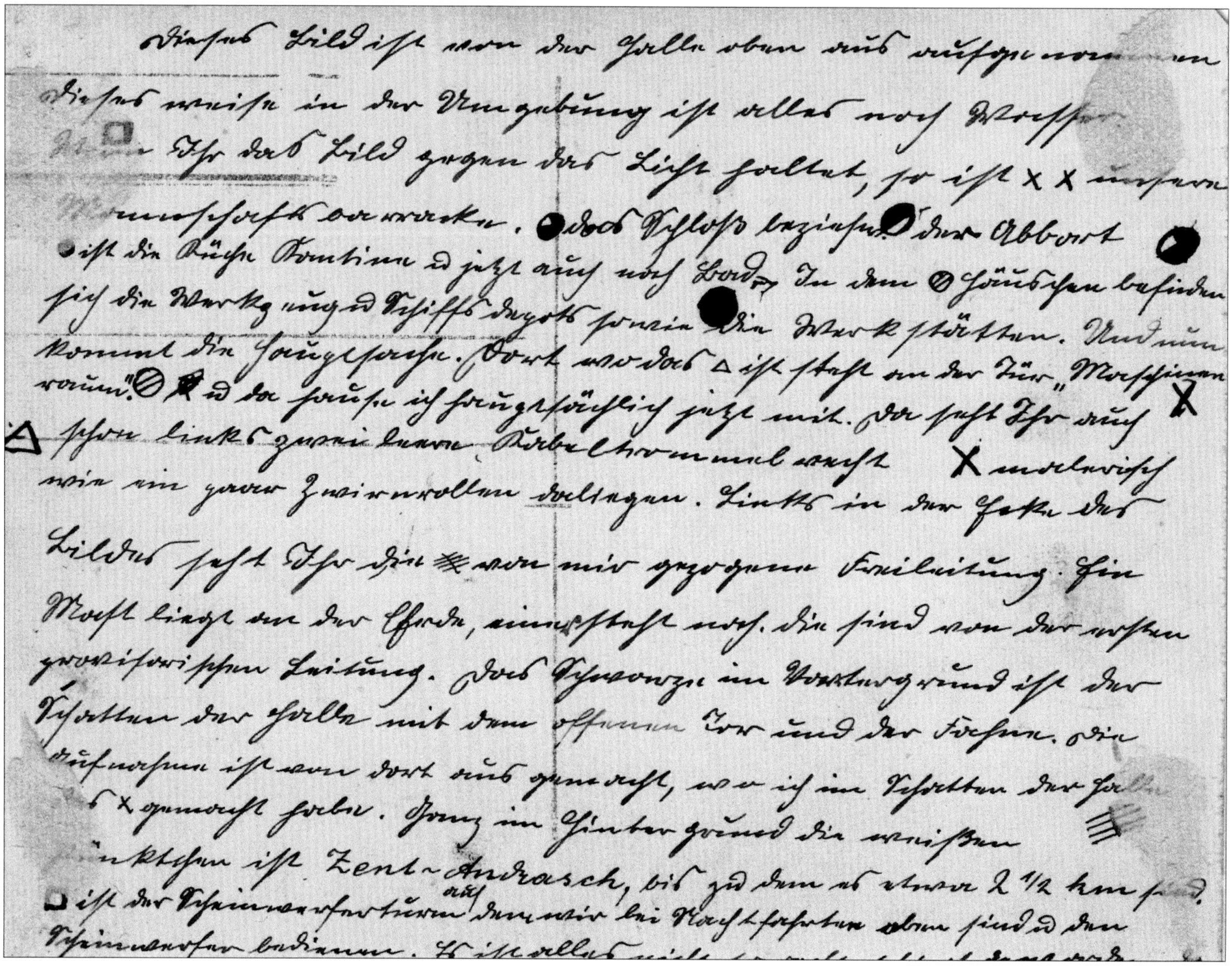

„Dieses Bild ist von der Halle oben ausgenommen – Dieses weiße in der Umgebung ist alles noch Wasser.
Wenn Ihr das Bild gegen das Licht haltet, so ist XX unsere Mannschaftsbaracke.
◕ das Schloß beziehen ▮ der Abbort ● die Küche, Kantine und jetzt auch noch Bad
⊘ In dem Häuschen befinden sich die Werkzeug- und Schiffsdepots sowie die Werkstätten. Und nun kommt die Hauptsache. Dort wo das △ ist, steht an der Tür „Maschinenraum“ ◎ und da (prüfe) ich hauptsächlich jetzt mit. Da seht Ihr auch schon links zwei leere Kabeltrommeln recht malerisch wie ein paar Zwirnrollen daliegen. Links in der Ecke des Bildes seht Ihr die von mir gezogene Freileitung. Ein Mast liegt auf der Erde, einer steht noch, die sind von der ersten provisorischen Leitung.
Das Schwarze im Vordergrund ist der Schatten der Halle mit dem offenen Tor und der Fahne, die Aufnahme ist von dort aus gemacht, wo ich im Schatten der Halle das X gemacht habe. Ganz im Hintergrund die weißen Pünktchen ist Zent-Andrasch, bis zu dem es etwa 2 ½ km sind.
□ Ist der Scheinwerferturm auf dem wir bei Nachtfahrten oben sind und den Scheinwerfer bedienen. Es ist alles nicht so recht scharf geworden. Da …

Image 73 Above: Text on the reverse side of image 71.

Image 74 Above: Accommodations, workshops, and warehouses across from the airship hangar (see scharekf.hu).

Image 75 Above: View of the gas works. The airship hangar is pictured in the background (image: Biber Archive).

to the airship hangar in order for other airship base operational requirements to be fulfilled.

Construction work for the airship hangar was complete in late October. The finished hangar measured 180 m in length, 55 m in width, and 35 m in height. The rapid completion of the construction project was made possible by the Austrian-Hungarian Wehrmacht, "making metal workers, lathe operators, and other iron industry specialists available for the task" (see SANKTANDRES/ 1981, page 165).

"Alongside the hangar, rooms were designated for administrative, personnel, and storage purposes, as well as a variety of workshops (joinery, sheet metal shop, forge, etc.). Lastly, additional utility spaces, cafeterias, and two large pools were added. The water necessary was supplied by the four artesian wells.

The railway track branch ended in front of a depot containing the chemicals necessary for gas production. An underground line led from there to the engine room, where the Zeppelin airships' gas cells and balloons were filled. Spare parts were kept in another storage space; there were enough spare parts to rebuild an entire airship in case of an emergency. A generator-driven engine supplied the necessary electrical power. Each room was equipped with fire extinguishers. North of the hangar, bombs were stored under lock in a fortified cellar (see SANKTANDRES/ 1981, page 165).

The aerial photograph (see image 70) shows how much progress had been made in the Timisoara airship base construction project; within a short period of time, multiple buildings (workshops and accommodation buildings), as well as transport routes (see left side of the airship hangar), railway

Image 76 Above: Overall view of Szentandrás airship base near Timisoara, code name "ADEBAR", including airship hangar, gas works on the left, towers for flashing lights and spotlights (left and right of the airship hangar), workshops, and storage warehouses (right), as well as its swampy surrounding areas (image: Biber Archive).

Image 77a Above: German wireless telegraph station, exterior view (see scharekf.hu).

connections, and material warehouses (right side of the airship hangar) were developed.

Images 71 through 75 provide additional details on the Timisoara airship base installation.

Not only did the machine room house an E-engine and a generator but also a converter and transformer station (see images 70 and 71). In order for the base to supply itself with gas, a gas works was under construction. Two pools from four artesian wells were utilized for gas production. Furthermore, a storage shed for gas bottles was established featuring high pressure and dispenser lines.

The view from the airship hangar (see image 71) as well as image 76 indicate that the grounds surrounding the airship base appeared to be slightly

Image 77b Right: A small hut and a tower formed part of the wireless telegraph and radio station at Timisoara airship base (image: Fabry Archive).

Image 78 Left: German wireless telegraph station, interior view (see scharekf.hu).

Image 78b Left: Interior of the radio station

Image 81 Facing Page, Below: Photograph of Paul Biber (2nd from right) setting up workshops for the operation of the airship base with his comrades on November 11, 1915. A note on the reverse of the image reads: "Electricians (Flintendrücker) at work" (image: Biber Archive).

Image 79 Left: View from the airship base premises across the swampy grounds to the village of Szentandrás (image: Biber Archive).

Image 80 Above: Farm visit in the town of Szentandrás (image: Biber Archive).

Image 82 Above: Paul Biber in his quarters with his comrades (image: Biber Archive).

watered-down and swampy.

In order to provide a reference point for airships in all environmental conditions, towers were established for flashing lights and spotlights (see image 76). In addition, the airship base was supplied with a "Drachenstation" where "measurements of meteorological data (air pressure, wind direction and speed, electrical atmospheric charge, etc.)" (see SANKTANDRES/ 1981, page 166) were recorded. This data was of high importance for the briefings held prior to an airship mission.

Furthermore, a German wireless telegraph station (see images 77a, 77b, 78a, and 78b) was established in the vicinity of the airship hangar, along with the Austrian-Hungarian Marconistation for the balloon department in the immediate vicinity of the village (see SANKTANDRES/ 1981, page 166).

As a whole, Timisoara airship base was put

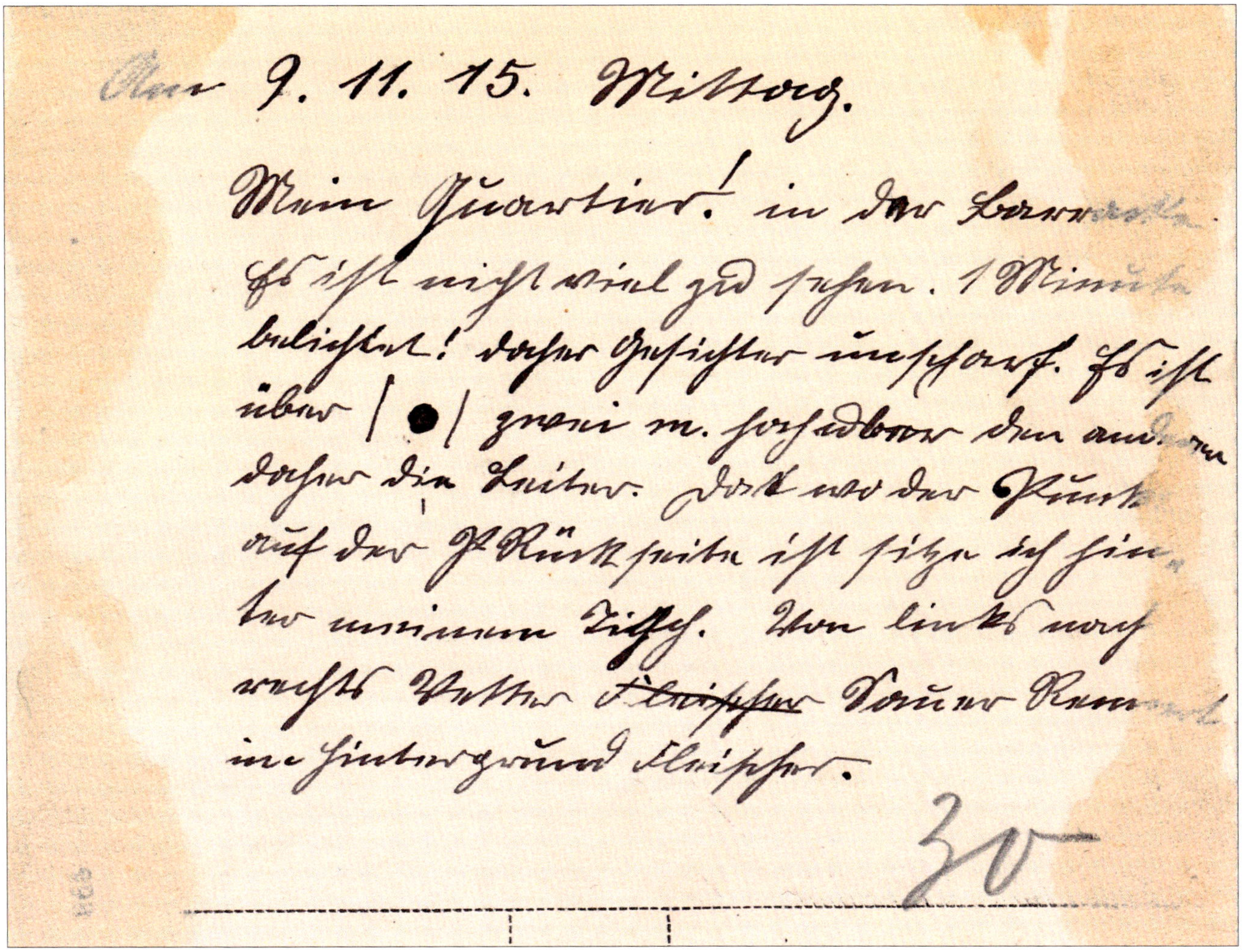

Image 83 Above: Reverse of photograph pictured in image 82 (image: Biber Archive).

„9.11.15, Mittag

Mein Quartier! In der Baracke. Es ist nicht viel zu sehen. 1 Minute belichtet! Daher Gesichter unscharf. Es ist über zwei m hoch über dem der anderen, daher die Leiter. Ich sitze hinter meinem Tisch. Von links nach rechts Vetter, Sauer, Rennert und im Hintergrund Fleischer.“

Image 84 Above: Text on the reverse of photograph pictured in image 82 (image: Biber Archive).

into operation in late October, and its operational readiness was established.

The village of Szentandrás (see image 79) was situated at a distance from the airship base. From time to time, Luftschiffer paid a visit to the village (see image 80). When traveling to the village or other locations nearby, the likelihood for muddy travels was high depending on weather conditions. This can be guessed by looking at image 76.

The airship base crew was usually busy setting up accommodations and workspaces (see images 81-84), as Timisoara airship base was soon to be ready to accept an airship.

4. From November 1915 – "LZ 81" Deploys to Timisoara Airship Base

Image 85 Above: Zeppelin LZ 81, airborne (image: Biber Archive).

Image 86 Below: Zeppelin "LZ 81" data before and after its modification.

Daten von Zeppelin (LZ 51) – militärische Bezeichnung „LZ 81":

- Typenbezeichnung:	LZ 55, Ausführung p		
- Bauart:	Starrluftschiff mit zwei Gondeln		
- Länge:	163,5 m	(nach Umbau 178,5 m);	
- Größter Durchmesser:	18,7 m;		
- Volumen:	32.000 m³	(nach Umbau 35.800 m³);	
- Nutzlast:	15.580	(nach Umbau 17.500 kg;	
- Zellenanzahl	16		
- Max. Steighöhe:	3.900 m		
- Reichweite:	4.300 km		
- Triebwerke:	4 x Maybach HSLu		
- Leistung:	4 x 240 PS (170 kW)		= 960 PS

(siehe: Griehl, 2010, S. 78)

4.1 Information on "LZ 81" and Its Transfer from Dresden

Zeppelin LZ 51 was part of the 32,000 m³ airship "p" construction series; of the 21 total airships in the series, the Imperial Navy received 11 (L10-L19, and L25) and the Army administration received 10.

The airship bearing construction number LZ 51 was built at "Löwenthal" shipyard in Friedrichshafen. Within the Army, Zeppelin LZ 51 was issued Army airship number "LZ 81".

During WWI, modifications were carried out on airships delivered to the Army (see SZ, 2010, page 19). In June and July of 1916, "LZ 81" was modified at Dresden, changing the type from "p" to "q" (for technical data, **see image 86**).

In favorable weather conditions, "LZ 81" **(see**

Image 87 Above: "LZ 81" airship nacelle with machine gun (image: see Post LZ 81).

Image 88 Right: Zeppelin "LZ 81" Kommandant Hauptmann Jacobi (image: Biber Archive).

image 85) was capable of reaching speeds up to 100 km/h. Two nacelles at the bottom of "LZ 81" could accommodate 12-14 crew members. Both nacelles were equipped with two machine guns each **(see image 87).**

"LZ 81" had its maiden voyage on October 7, 1915 under the command of Hauptmann Jacobi **(see image 88).** The airship was based in Dresden from October 9, 1915.

As a result of the rapid commissioning of Szentandrás airship base near Timisoara by Königl. Sächs. Feldtrupp für Luftschiffe Nr. 14 ("LT 14"), Army airship "LZ 81" was ready to begin its flight from Dresden to Timisoara on November 1, 1915 under the command of Hauptmann Felix Jacobi.

K.u.k. chief officers Franz Mannsbarth and Siegfried Heller (who was commanded to Germany in mid-October 1915) were part of the crew during the ferry flight. The airship made an intermediate stop at Teschen (see Kustan & Penz, pages 15/16).

Airship "LZ 81" subsequently continued its journey to Temesvár at a distance of 1.088 km. As a

Image 89 Above: Paul Biber standing on the roof of the airship hangar near the doors on November 2, 1915; his eyes searching the horizon for airship "LZ 81" which had announced its imminent arrival at Szentandrás-Temesvár (image: Biber Archive).

result of the rapid ferry flight of the airship, a swift operation against Allied Forces on the Balkans front was secured.

In Timisoara, airship base personnel stood on top of elevated structures (airship hangar, towers for leading lights) in order to look out for the Zeppelin (see image 89). Paul Biber courageously climbed the roof of the airship hangar near the doors (see image 89) on November 2, 1915. His view included the swampy area surrounding the airship base. He had little interest in the muddy grounds around him however, as his only focus was to look for Zeppelin "LZ 81".

Thanks to the telegraph station, the airship was able to announce its impending arrival at Szentandrás-Temesvár. Then, finally, a small dot on the horizon slowly started increasing in size, prompting "Landungstrupp" crew members (Haltemannschaft) to begin carrying out their preparation tasks.

On November 2, 1915 at 1400, "LZ 81" was the first German airship to land at Szentandrás-Temeswar airship base (Hungary/present day Romania). "An hour prior to the airship's arrival, the news rapidly spread through town. Everyone who was able to witness the sight (particularly school-aged children, teachers, and local authorities) assembled on the 'large bridge', carefully watching the airship that majestically approached, circling above the hangar multiple times before proceeding towards Kowatschi, Tschernetzhausen, and the Sanktandres fair; it paused for quite some time overhead the cemetery as if to commemorate the dead. Finally, after approximately 30 minutes, the airship arrived on the tarmac in front of the hangar. It was a beautiful fall day, the designator LZ 51 (81) was easily discernable. While Germans and Romanians welcomed the giant flying object with a standing ovation, the gypsy women in attendance cried and crossed themselves repeatedly; herding geese scattered, fleeing towards the town. Many Sanktandres citizens vividly remember this day." (SANKTANDRES, 1981, page 166).

Airship LZ 81 was secured on the tarmac in

Image 90 Above: Securing LZ 81 and welcoming of the crew (image: Racil Collection).

Image 91 Below: The crew of LZ 81 under the command of Hauptmann Felix Jacobi (image: Racil Collection).

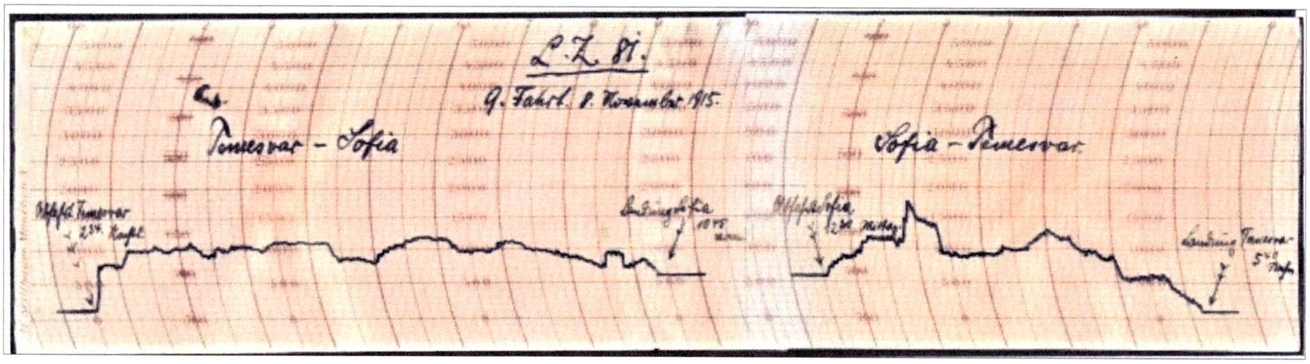

Image 92 Above: “LZ 81” flight barogram to Sofia (see scharekf.hu)

Image 93 Below: Zeppelin “LZ 81” arriving at Sofia airport, transporting Duke Johann Albrecht von Mecklenburg (image: schrekf.hu and Post LZ 81).

front of the airship hangar while the crew was welcomed **(see image 90).** The crew of LZ 81 under the command of Hauptmann Felix Jacobi presented themselves for a photograph in a slack formation **(see image 91).**

The crew of airship LZ 51 (81) counted “12 men, including erster Schiffsführer (commander) Felix Jacobi, zweiter Leutnant Barth, Funkeroffizier (radio operator) Nikisch von Rosenegg, technischer Ingenieur (technical engineer) Maukee, and Obersteuermann (navigator) Jireker. Bobby the dog was also part of the crew, he served as a mascot.” (SANKTANDRES, 1981, page 166).

The Austrian-Hungarian ground personnel held a welcoming feast with the following menu items: hydrogen soup (cauliflower soup), poultry with motor oil sauce (tartar sauce), Reuter-Havas-roast (duck roast), Zeppelin pie (a replica of LZ 51), Zeppelin bombs (ice cream) with hand grenades and arrows (wafer rolls), fruit, Turkish coffee, Zeppelin

Image 94 Right: Tsar (king) Ferdinand I. of Bulgaria in negotiations with the Duke of Mecklenburg after the arrival of airship LZ 81 in Sofia on November 9, 1915 (image: see Post LZ 81).

cigars, and Werschetz wine.

The Sanktandres municipal council hosted another welcoming reception for the crew at Rausch restaurant. The meal served was chicken "Paprikasch". The innkeeper used a particularly "hot" kind of Paprika. Stemper Heinrich (1890) still remembers the following amusing incident: As the guest of honor, Kommandant Jacobi was to take the first bite of the dish. Horrified, he jumped up and screamed, thinking he was about to be poisoned.

He finally calmed down after each member of the municipal council served themselves without any hesitation and proceeded to eat their meals. It was not until the Bakowa wine was served Jacobi realized that there could not have been a better foundation. The high spirits were the best proof" (SANKTANDRES, 1981, page 166/167).

Between November 2, 1915 and January 19, 1916, the operational area of "LZ 81" was at the

Image 95 Above: View of the bustling railway station in Timisoara (image: Biber Archive).

Image 96 Above: "In front of Timisoara railway station. Aren't the two men in the corners on the left a wonderful sight?" (see reverse of image 95) (image: Biber Archive).

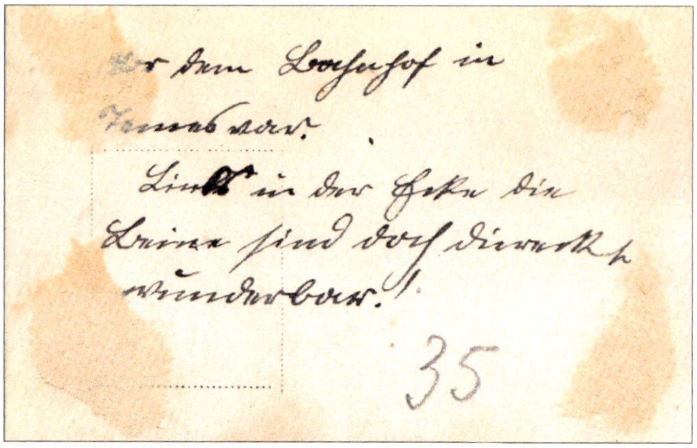

Image 97 Above: Reverse of image 96 captioned "In front of Timisoara railway station" (Biber Archive).

Image 98 Above: Paul Biber in Timisoara cathedral square in front of a Roman cathedral (Romano-Catolica) built during the baroque period (image: Biber Archive).

southeastern front. At the time, there was no direct land connection between Austria-Hungary and allied Bulgaria. On November 8, 1915 "LZ 81" transported a commission led by duke Johann Albrecht von Mecklenburg, special ambassador to the German emperor from Timisoara to Sofia (Bulgaria) for personal negotiations with Ferdinand I., tsar (king) of Bulgaria.

On its way to Sofia, airship "LZ 81" crossed overhead Serbia **(see image 92).** "LZ 81" touched down safely in Sofia **(see image 93).** One of the passengers onboard the German Zeppelin was duke Johann Albrecht von Mecklenburg himself. Tsar (king) Ferdinand I. of Bulgaria welcomed the duke of Mecklenburg after airship LZ 81 arrived at the airport on November 9, 1915 **(see image 94).**

Image 99 Above: German soldiers and Romanian farmers (image: Biber Archive).

Image 100 Below: Siebenbürginnen offering traditional homemade lace fabrics for sale in Timisoara (image: Biber Archive).

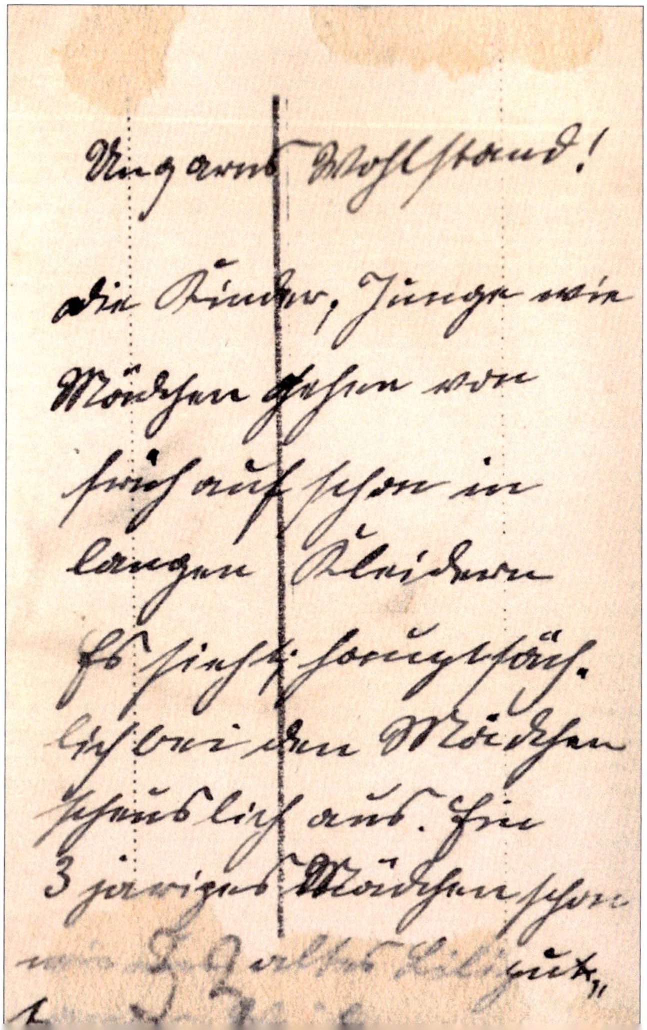
Ungarns Wohlstand!

die Kinder, Junge wie Mädchen gehen von früh auf schon in langen Kleidern so sieht hauptsächlich bei den Mädchen scheußlich aus. Die 3 jährigen Mädchen schon

Image 101 Facing Page Above: Holzsäger" (wood cutters) at work on the street in front of Timisoara "CENTRALBANK" (image: Biber Archive).

Image 102 Facing Page Below Left: "Strong boy" – boy holding a large piece of wood (image: Biber Archive).

Image 103 Facing Page Below Right: Reverse of photograph in image 102. (image: Biber Archive).

Image 104 Above: Fruit salesperson in Timisoara (image: Biber Archive).

Following the conclusion of negotiations (Bulgaria committed itself to entering the war in support of the Central Powers), the duke returned to Timisoara on "LZ 81" (see Kustan & Penz, page 16).

4.2 Exploring the Surroundings of Timisoara Airship Base

Following the successful commissioning of the airship base, occasional joint excursions were made to Timisoara and the surrounding areas **(see images 95 through 105).**

At the time, Timisoara was part of Austria-Hungary, a lively trading town with several public parks, well-constructed roads, and impressive buildings. Airship base personnel spent time people watching near the railway station and taking in the beautiful public buildings such as the town hall, a variety of churches, and fancy hotels. They also attempted to communicate with the locals, especially German immigrants.

Paul Biber felt compelled to capture an image of two "wonderful" men on the corner of the entry door to Timisoara railway station (see images 96 and 97).

On the reverse of the image captioned "Wood cutters at work" (see image 101), Paul Biber made the following note regarding the open hatch through which the cut wood was deposited into the basement of the house: "If you ever ended up breaking your leg on one of these open hatches in the darkness of the night, it would be a very painful experience. If only people were more concerned about this issue and would shut the hatch after they finished working for

Image 105 Above: Soldiers negotiating with liquor salespeople in Timisoara (image: Biber Archive).

the day." The image captioned "Strong boy…" (see image 102) also has a note on the reverse (see image 103). It reads: "Hungary's wealth- children, boys as well as girls, wear long dresses at an early age. It is a terrible sight to behold, particularly in girls. A 3-year-old girl resembles an aging Lilliput."

4.3 "LT 14" Practices Normal Operations with Airship "LZ 81"

In order for airships to land, be maintained, refueled, and depart again, many tasks had to be carried out. As the airships were utilized in a war zone, they were frequently damaged. If they reached the safety of an airship base, they were subject to extensive repairs. Carrying out these various necessary tasks required a large number of skilled personnel.

Following an in-range announcement of "LZ 81", a variety of preparations were carried out immediately. For example, landing devices needed to be brought outside. The numbers (see image 106) were used for the display of temperatures.

Ground personnel often had to run to meet the airship (see image 107), attempting to grab hold of the ropes that were dropped by the crew from the nacelles in order to secure the vessel to the ground. Soon after, orders were given to move the airship to the hangar ("Luftschiff-zur Halle marsch!"). The airship was slowly pulled downward and maneuvered toward the tracks. Using the tracks, the airship was then guided toward the hangar. Afterwards, the airship was pulled into the hangar

Image 106 Above: Preparation work (temperature display) for the landing of an airship (image: Biber Archive).

("Einhallen") and once inside, finally turned into the correct position (see image 108) to be securely anchored to the tracks using retaining ropes.

In preparation for a scheduled airship departure, base personnel filled it with hydrogen in a timely manner. For this task, gas works tanks located by the artesian wells located adjacent to the hangar were utilized.

Another task involved equipping the Army airship with ammunition as required in accordance with orders for the particular mission. Only trained airship base personnel were allowed to carry out this task.

If "LZ 81" was inside the hangar when orders to move it outside were issued ("Aushallen"), Army airship "LZ 81" crew members gathered for

Image 107 Above: Airship ground personnel walking (frequently running) to meet the airship and grab hold of the ropes to help secure it to the ground (image: Biber Archive).

Image 108 Above: Zeppelin "LZ 81" positioned inside the Timisoara airship hangar, anchored to the tracks with ropes (image: Biber Archive).

a briefing of the plan of attack (see images 109 and 110).

Luftschifftrupp 14 team members gathered to receive their task assignments. Kamerad Noack, for example was tasked with cleaning the right-hand track. Several team members opened the hangar doors while the majority of the ground crew positioned themselves near the ropes.

The crew of "LZ 81" took up their positions inside the airship nacelles. The airship Kommandant reviewed the latest weather information, focusing on wind development (direction, speed/max current velocity). Possible execution strategies and flight routes to the intended destination were discussed.

Hauptmann Karl Lumbe Edler von Mollowitz, Kommandant of "ADEBAR" airship base, issued orders to ground personnel for safely removing the airship from the hangar after consulting with Army airship "LZ 81" Kommandant Hauptmann Felix Jacobi. Thereafter, the airship was brought into position for the departure and ascent sequence (see images 111-114). In preparation of moving the airship out of the hangar, ropes from "LZ 81" were attached to both tracks (see image 111). Soon after, the order "Aushallen!" (move airship out of the hangar) was given. This command prompted

Image 109 Above: LZ 81 crew including Kommandant Hauptmann Felix Jacobi (seated, 3rd from left) during their duty assignment at Timisoara airship base (image: Biber Archive).

"Landungstrupp" personnel to grab hold of the ropes and pull the airship out of the hangar. Once outside, "Landungstrupp" (Haltemannschaft) members continued to hold on to the ropes while keeping the airship securely attached to the tracks and moving it into the correct position for departure.

A large number of ground personnel was necessary to move the airship, change its position, and to keep it balanced, particularly in windy conditions. The many ground crew members maneuvered the airship using ropes. The airship needed to be maneuvered into the wind for departure.

Image 114 clearly shows the airship positioned at a 90-degree angle to the tracks. This indicates that it was probably pulled from the center of the hangar and repositioned into the wind once outside or moved sideways in order to reach its ideal position for departure. Ground crew ("Haltemannschaft") members used skills and effort in order to maneuver airship "LZ 81" into the proper position for departure.

Once the desired departure position was reached (see image 115), the Haltemannschaft awaited the ascent command (see images 116 and 117). The command was "Luftschiff hoch!" Slowly, the

Image 110 Above: Seal of LUFTSCHIFF-KOMMANDO L.Z. 81. (see Post LZ 81)

Image 111 Left: Kamerad Noack cleaning the right-hand track in preparation for the airship's departure (image: Biber Archive).

airship gained altitude (see image 118). Soon after, it disappeared from view of the base (see image 119a and 119b).

The departure maneuver (as well as the landing maneuver) bore several risks for the ground crew. It was possible for the wind to cause a sudden rise of the airship. If a ground crew member did not let go of their rope in time, the airship pulled them up into the air. In this case, hydrogen was immediately released in order to cause the airship to descend back to the ground. The departure and landing maneuvers required high concentration, quick actions, as well as experienced ground personnel and airship crews.

4.4. December 1915- January 1916 – Flights on "LZ 81" Revealed Issues with Drive Motors

Good food supply contributed significantly to the morale of the crews at Timisoara airship base. Therefore, crew members routinely checked the

Image 112 Above: Gathering ground personnel on the ropes of "LZ 81" in preparation to move the airship out of the hangar ("Aushallen"); ropes were attached to both tracks (image: Biber Archive).

Image 113 Above: Moving "LZ 81" out of the hangar. Timisoara, November 1915 (image: Biber Archive).

Image 114 Above: After being moved out of the Timisoara airship hangar in November 1915 ("Aushallen"), the "Haltemannschaft" maneuvered "LZ 81" into a favorable position for departure (image: Biber Archive).

Image 115 Above: The airship reached its position for departure (image: Racil Collection).

Image 116 Left: The LT 14 Haltemannschaft secured LZ 81 in a safe position, awaiting orders for departure (image: Racil Sammlung)

Image 117 Above: Landungstrupp members firmly holding on to ropes and nacelle handlebars, awaiting the "Luftschiff hoch!" command, prompting them to release the ropes and handlebars (image: see scharekf.hu)

general farming status while in town, focusing specifically on sufficient amounts of animals and vegetables. This pig (see image 120) may not have been aware of its imminent end of life.

Individual ground crew accommodations were equipped with small kitchens where food was prepared (see image 121). Hungry crew members arrived, waiting for their metal bowls to be filled (see image 122).

It is probable that the baby calf in the photograph (see image 123) stayed alive for one or several more years.

Image 118 Above: The "LZ 81" in November 1915 following a successful departure and ascent (image: Biber Archive).

Image 119b Inset: Aerial view of Timisoara airship base including airship hangar, barracks, and rail connection. (Photo Archiv Fabry)

Image 119A Above: The "LZ 81" at high altitude overhead Timisoara airship base.

Time passed quickly, and soon it was Christmas. Everyone was intent on creating a holiday atmosphere. Luftschifftrupp "LT 14" electrical engineers made themselves as comfortable in their workshops as circumstances permitted. On the reverse of the Christmas photograph **(see image 124),** Paul Biber made the following note: "This is to show mother that the fur is still intact. I took it off inside the warm room but soon after, Comrade Schlesinger grabbed it." (image: Biber Archive).

Image 120 Above: A comrade riding a woolly pig. (image: Biber Archive)

Image 121 Above: The "LZ 81" in November 1915 following a successful departure and ascent .(image: Biber Archive)

Image 122 Above: Soldiers waiting for food being handed out at the cafeteria (image: Biber Archive).

Image 123 Left: Soldiers and village locals pose with a young calf (image: Biber Archive).

Image 124 Above: Christmas celebration 1915 at Szentandrás-Temeswar airship base electrical workshop/engine room with a note on the reverse of the image.

A Christmas tree was decorated, and contents of packages mailed from home were shared among comrades. Photographs taken of the Christmas tree and personnel (see images 124 and 125) were developed and sent to relatives at home (see images 126 and 127).

The crew of airship "LZ 81" celebrated Christmas in the hangar. Tables were arranged, little gifts handed out to "LZ 81" crew members, a tree was put up, and everything was decorated for the joint holiday dinner (see image 128).

Each aeronaut received a large Christmas greeting card as a souvenir from the joint 1915 Christmas celebration at Timisoara (see image 129).

Paul Biber utilized the reverse of a photograph taken of the Christmas celebration inside the engine room as a field postcard (see image 126). He briefly described the New Year's Eve event (see image 127). According to his account, Army airship "LZ 81" had its first war patrol flight in the Balkans on December 31, 1915. The mission goal was to bomb the city of Saloniki, particularly its port facility. The attack was canceled due to engine problems, however.

It should be noted that "LZ 81" had been equipped with new, performance-enhancing 240 HP Maybach engines. These engines had not previously been tested on missions at the front, however; particularly when crossing overhead mountainous regions carrying bomb loads. As a result, further war patrol flights in conjunction with crossings of the Macedonian mountain ridges were avoided. Such a crossing was "considered too dangerous, or impossible, respectively." (see Ploes, 2018, page 16).

Image 125 Above: Paul Biber (far left) at the Szentandrás-Temeswar airship base electrical workshop/engine room, Christmas 1915 (image: Biber Archive).

4.5 January 1916 – Transition to Airship Hydrogen Replenishment Self-sufficiency

Paul Biber was a precision engineer **(see image 130).** Prior to being drafted, he worked at Firma Rosenmüller in Dresden (where he produced instruments for flow measurement, for example anemometers and airspeed indicators). In preparation for his Meister- und Technikerprüfung (master's and engineer examination), he attended a variety of advanced courses at Städtische Gewerbeschule Dresden (Dresden municipal trade school), including "Dynamos, DC motors", "Telegraphy and Telephony", as well as "General Electrical Engineering" **(see image 131).** As a result of Paul Biber's extensive training in a related field, he was put in charge of maintaining the functionality of direct current production equipment at Timisoara airship base. This equipment was of paramount importance for the in-house hydrogen production at the "Gasanstalt" (gas works). The facility was located inside of a long barrack of relatively stable construction **(see image 132a). O**nly for a short time a provisional power generation was carried out by driving the generator by means of running LKW truck motors (see Image 132b).

Every morning at 3:00 am, Paul Biber reported for duty at the engine room to operate the equipment for DC production. The transformer station required continuous monitoring in order to assure that direct current would reach the gasworks on base via the distribution line extending along the road. The

Image 126 Right: Reverse side of photograph pictured in image 125 (Biber Archive).

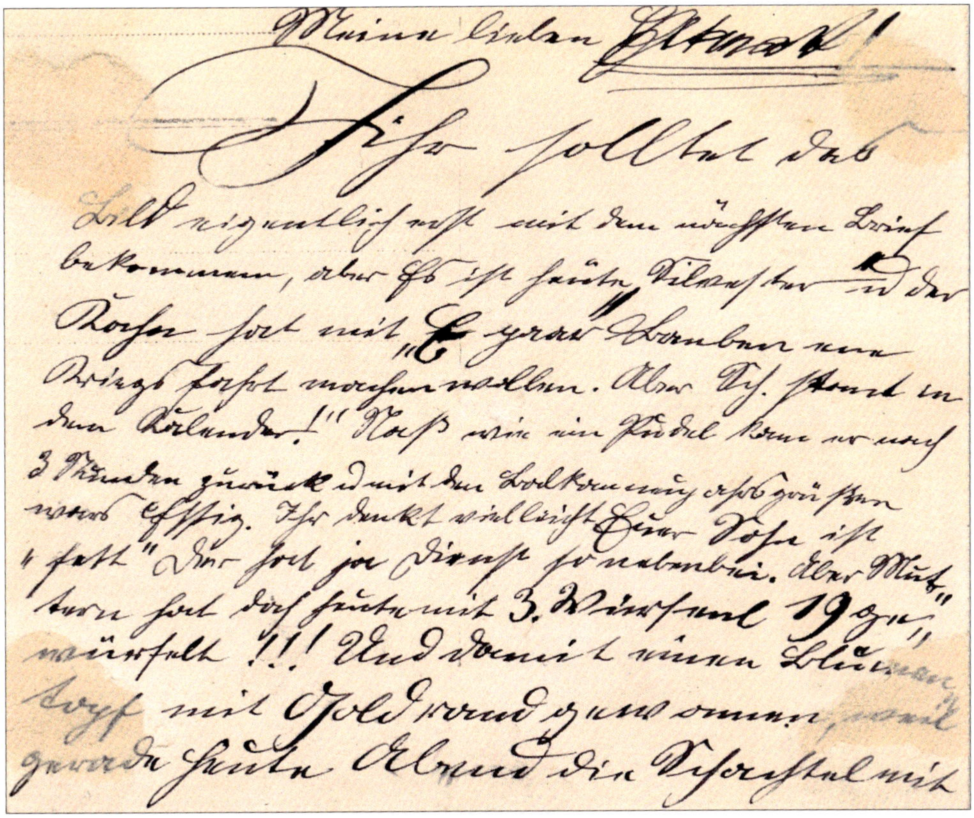

Image 127 Below: Text on reverse side of photograph (see image 126).

„Meine lieben Eltern

Ihr solltet das Bild eigentlich erst mit dem nächsten Brief bekommen, aber es ist heute Silvester und der ....hat mit „.......... nun Kriegsfahrt machen wollen. Aber nix stand in dem Kalender!" Naß wie ein Pudel kam er nach 3 Stunden zurück u. mit den ... grüßen wars .... Ihr denkt vielleicht Euer Sohn ist „fett" der hat ja Dienst so nebenbei. Aber Mutter hat doch heute mit 3 Würfeln 19 gewürfelt!!! Und damit einen Blumentopf mit Goldrand gewonnen. ..."

establishment of the distribution line was led and overseen by Paul Biber. The gas leaching was also carried out provisionally in the open at the beginning (see Image 132c). Supervisors were not comfortable entering the transformer room because of the big red lightning bolt with the inscription "mortal danger" on the door. As a result, Paul Biber was able to work largely on his own terms without frequent supervision.

Securing energy supply for hydrogen production was now Paul Biber's main area of responsibility (see images 133 through 137).

Paul Biber's notes on the reverse of photographs indicate his involvement in the machine works installation. He established the electrical long-distance pipeline for the transmission of produced direct current from the engine room to the gas works. Electricians settled comfortably into the transformer room for the duration of their deployment. Image 113 taken in March 1916 shows Paul Biber enjoying potatoes in bacon fat. The sausages hanging from the ceiling are a clear

Image 128 Above: "LZ 81" crew Christmas celebration in the Timisoara airship hangar (image: Biber Archive).

indication that he never had to go hungry.

In addition to his regular duties, Paul Biber helped carry out other tasks at the airship base, including assisting during "Ein- und Aushallen" (moving the airship into or out of the hangar), or making repairs to electrical and mechanical equipment.

There was a variety of electrical equipment on the airship base that needed to be installed, operated, and maintained. As a result, the "electricians" group substantially grew in size (see image 138).

4.6 January 1916 – Farewell, "LZ 81"

On January 15, 1916 Air Force Command ordered Hauptmann Hermann Kühn to take command of the "ADEBAR" airship detachment (see Kustan & Penz, page 17).

"LZ 81" did not remain in Timisoara for long. The flights in the Balkans area revealed to Commander Hauptmann Jacobi several shortcomings with the airships' drive motors. The failures were associated with lubrication; additionally, there were other faults related to engine run (see Navy, 1992, III, page 10).

On January 19, 1916 Timisoara ground personnel bid farewell to "LZ 81" as the airship departed for Dresden (see images 139 through 141). There is a message from airship squad field forces 14 Commander of Dresden-Kaditz airship base, dated January 22, 1916 indicating the safe arrival of airship LZ 81 at Dresden (see record Nr. 1033). At Dresden, the airship was equipped with different engines before it left for the western front on a different duty assignment.

The Austrian-Hungarian Army strived to acquire its own airships for utilization in the war. This required suitable commanders and well-trained personnel. Oblt. Max Macher was asked to submit a detailed report outlining the suitability of high-ranking officers. He assessed several officers and concluded his report by saying that it would be

Image 129 Above: Postcard on the occasion of the 1915 "LT 14" and "LZ 81" Christmas celebration in Timisoara (image: Biber Archive).

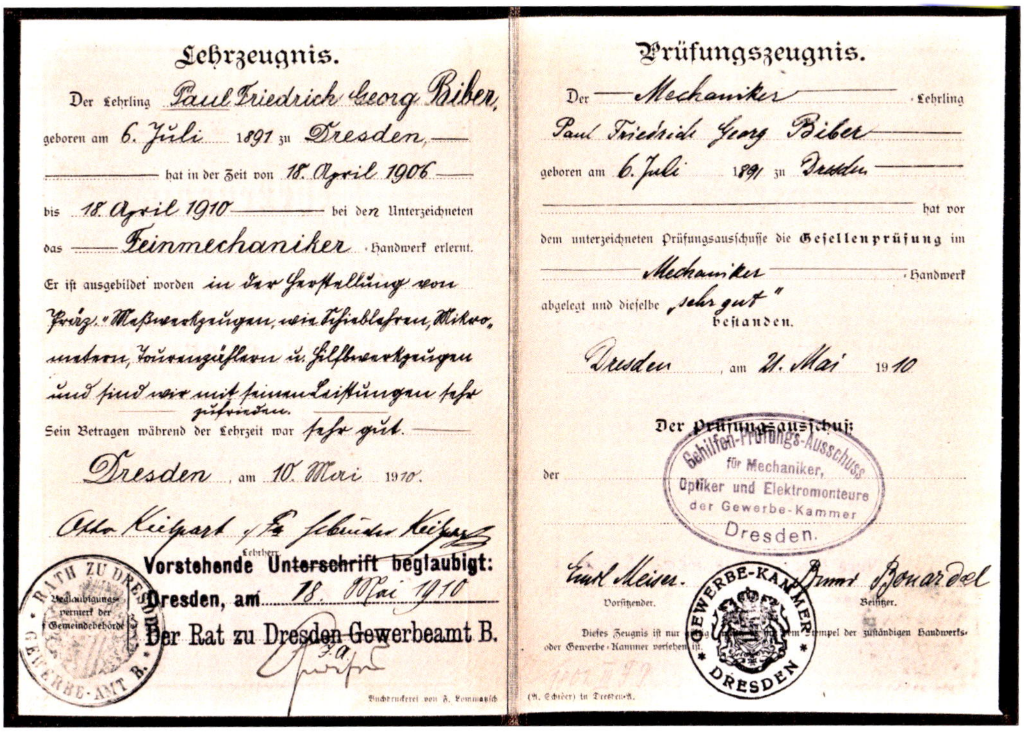

Lehrzeugnis.

Der Lehrling Paul Friedrich Georg Biber, geboren am 6. Juli 1891 zu Dresden, hat in der Zeit von 18. April 1906 bis 18. April 1910 bei den Unterzeichneten das Feinmechaniker-Handwerk erlernt.

Er ist ausgebildet worden in der Herstellung von Präz.-Maßwerkzeugen, wie Schieblehren, Mikrometern, Toleranzlehren u. Hilfswerkzeugen und sind wir mit seinen Leistungen sehr zufrieden.

Sein Betragen während der Lehrzeit war sehr gut.

Dresden, am 10. Mai 1910.

Lehrherr.

Vorstehende Unterschrift beglaubigt:
Dresden, am 18. Mai 1910
Der Rat zu Dresden Gewerbeamt B.

Rath zu Dresden · Beglaubigungsvermerk der Gemeindebehörde · Gewerbe-Amt B.

Buchdruckerei von F. Lommatzsch

Prüfungszeugnis.

Der Mechaniker-Lehrling Paul Friedrich Georg Biber geboren am 6. Juli 1891 zu Dresden hat vor dem unterzeichneten Prüfungsausschusse die Gesellenprüfung im Mechaniker-Handwerk abgelegt und dieselbe „sehr gut" bestanden.

Dresden, am 21. Mai 1910

Der Prüfungsausschuß:
Gehilfen-Prüfungs-Ausschuss für Mechaniker, Optiker und Elektromonteure der Gewerbe-Kammer Dresden.

Emil Meiser. Vorsitzender. Bonardel Beisitzer.

Gewerbe-Kammer Dresden

Dieses Zeugnis ist nur gültig, wenn es mit dem Stempel der zuständigen Handwerks- oder Gewerbe-Kammer versehen ist.

Image 130 Right: Excerpt from Paul Biber's examination certificate (image: Biber Archive).

Allg. Elektrotechnik	1	2a	Algebra	1	1b
Meßkunde (Elektr.)	1	2a	Geometrie	1	1b
Telegraphie u. Telephonie	1	1	Trigonometrie	1	1b
Dynamomaschinen; Gleichstrommotoren	1	1	Mathem. Übungen	1	1b

BEMERKUNGEN: Paul Biber erhielt Ostern 1909 eine schriftliche Belobigung, Ostern 1911 und Ostern 1913 eine Prämie.

Image 131 Facing Page Above: Excerpt from Paul Biber's school leaving certificate at Städtische Gewerbeschule zu Dresden (Dresden municipal trade school) (image: Biber Archive).

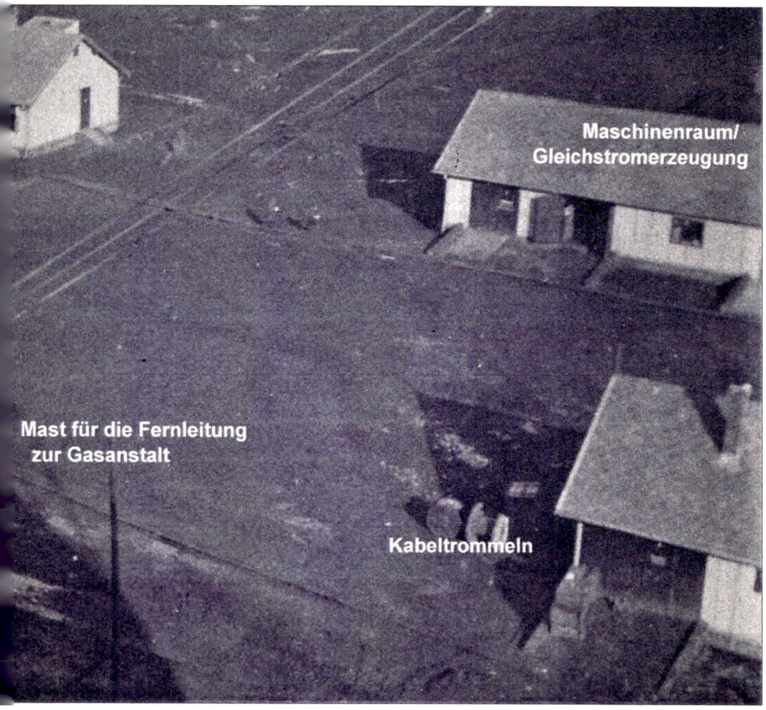

Image 132c Above: DC production inside the engine room using a long-distance pipeline for transmission (established by Paul Biber) leading to the gas works (image: Biber Archive).

Image 132b Above: Tentative means of power generation; the generator was powered by a running truck engine and a belt drive.

Image 132c Above: Tentative means of gas production for on-site supply and replenishment of hydrogen for the airship. Later, gas production and filling of gas cylinders took place inside a barracks.

Image 133 Above: Paul Biber and Kamerad Vetter having a meal in the transformer room. The following note was added to the reverse of the image: "March 1916 – inside the high-voltage room at Timisoara base. We were eating potatoes in bacon fat." (image: Biber Archive)

in the interests of the cause to have one or two additional officers sent to Germany for training so as to have a greater selection at a later time.

Those officers who were not selected to command an airship would soon be able to find positions in different Air Force departments based on the new skills and knowledge acquired during their training in Germany.

Mid-January 1916 the introduction of airships with the k.u.k. Army was discontinued. Oblt. Max Macher remained in Germany in order to provide continuous information regarding new features of German airships as well as general tests and trials whose outcome might have an effect on the Air Force to k.u.k. Air Force troops (see Kustan & Penz, pages 16-18).

On January 27, 1916 the German Emperor's birthday was formally celebrated at the Timisoara airship hangar with a roll call by ground personnel and officers (see image 142).

Following the roll call, a joint celebration was held (see images 143 and 144).

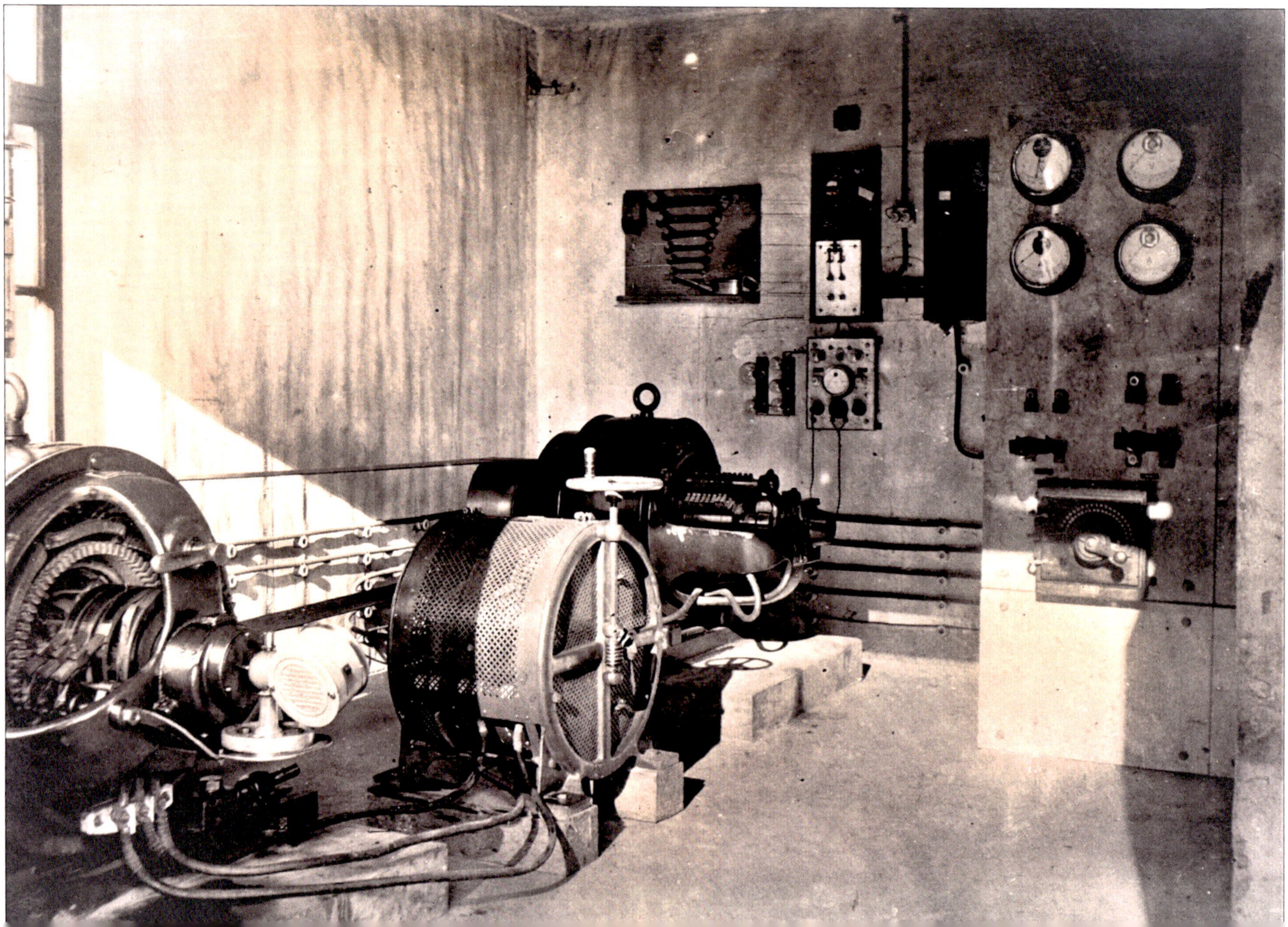

Image 134 Facing Page Above: Converter room with transformer stand (Biber). The following note was made on the reverse side of the photograph: "February 12, 1916 – inside the high-voltage room; we were well set up for the duration of our stay." (image: Biber Archive)

Image 135 Facing Page Below: Engine room for DC generation, including transformer stand, Timisoara (image: Biber Archive).

Image 136 Above: Paul Biber inside the engine room at Timisoara airship base (image: Biber Archive).

Image 137 Right: Transfer station for the high-voltage transmission line established by Paul Biber (image: Biber Archive).

LZ 84

Image 138 Facing Page Above: Paul Biber (2nd from left) with his electricians, midday on January 14, 1916 (image: Biber Archive).

Image 139 Facing Page Below: "LZ 81" being moved out of the hangar for its return flight to Dresden in January 1916 (image: Biber Archive).

Image 140 Right: Kommandant Felix Jacobi says goodbye to LZ 81 prior to its departure (image: Racil Collection).

Image 141 Below: Airship LZ 81 is ready to depart for Dresden (image: Racil Collection).

Kaiser-Geburtstagsfeier in Temesvar. 27. 1. 1916

Image 142 Facing Page Above: Roll call on the occasion of the imperial birthday celebration at Timisoara (image: Biber Archive).

Image 143 Facing Page Below Left: Paul Biber (2nd from left) attending the imperial birthday celebration with his "LT 14" comrades (image: Biber Archive).

Image 144 Above: Paul Biber attending the imperial birthday celebration with his "LT 14" comrades (image: Biber Archive).

5. War Patrol Flights on "LZ 85" Departing Timisoara Airship Base from February 1916

5.1 "LZ 85" Arrival at the Airship Base in January and Its First Bomb Attack Flight

On September 12, 1915 Kommandeur Oberleutnant Ernst Scherzer took command of airship "LZ 85" in Potsdam. The airship was stationed at Alleinstein airship base beginning September 26, 1915. From there, Oberleutnant Ernst Scherzer departed on three successful war patrol flights to Latvia onboard "LZ 85" in October, bombing Dinaburg, Riga, and Minsk (see Navy, 1992, III, page 11).

Hauptmann Scherzer led "Sachsen" during Hindenburg's summer military campaign against Russia after Luftschiffkapitän Lehmann. His First Officer, Leutnant Nippe proved to be a very successful crew member as "Sachsen" Bombenoffizer over Belgium (see Lehmann, 2019).

Mid-January, Ernst Scherzer transferred to Timisoara with "LZ 85". He arrived at Timisoara on schedule on January 27.

After remaining in his previous position for less than a month, Kommandeur Hauptmann Kühn joined Luftschiffer-Ersatzbataillon in Vienna. On February 10, 1916 Hauptmann Kornél Poppe (see image 145) took command of Luftschifferdetachement "ADEBAR" (see Kustan & Penz, page 18).

Hauptmann Kornél Poppe considered himself particularly called to lead LFT as he had been part of the Navy airship detachment at Nordholz near Cuxhaven for quite some time, having acquired all the relevant expertise and technical knowledge for the position.

The higher registration number of "LZ 85" compared to "LZ 81" indicates the date of manufacture of "LZ 85" is more recent. This is not the case, however – the maiden voyage of "LZ 85" took place on September 12, 1985 whereas the first trip carried out with "LZ 81" was on October 7, 1915. The most important aspect of these dates relates to the airships' operational capability in the Balkans area, as LZ 85 had been fitted with the reliable 210 HP Maybach engines (see Ploes, 2018, page 17).

Image 145 Above: Hauptmann Kornél Poppe –Szentandrás airship base commander. (Temeswar/image: Biber Archive)

"Officers praised the solid construction of the airship; it was very stable, a considerable different from LZ 81. Lt. Nippe was impressed with the crew which consisted of carefully selected members. During its deployment to Timisoara, airship LZ 85

Image 146 Right: Hauptmann Scherzer (middle) with Oberleutnant Nippe (left) and an officer from Timisoara airship base (image: see scharekf.hu)

carried out three attacks on Saloniki." (Navy, 1992, III, page 11).

Following drastic weather improvements in the Timisoara area in late January, "LZ 85" carried out its first war patrol flight in the Balkans (see image 147) from Timisoara which took place between January 31 and February 1 under the command of Hauptmann Ernst Scherzer (see image 146).

"LZ 85" departed Timisoara at 1525 on a mission to bomb Saloniki. The trip there would not be without its challenges (see image 148). The airship had to cover a considerable distance, leading across a high mountain range causing severe turbulence and required the airship to climb to higher altitudes. The heavy weight of the bombs onboard proved to be a challenge for the engines which in turn forced the crew to carefully manage their fuel supplies.

"The airship flew on course toward Versecz and thereafter towards Morawa valley...at 2025, they reached Nisch at an altitude of 2000m...during the continuation of the flight it was discovered that the elevation numbers recorded on the charts were much too low. At 2105, the crew lost sight of the ground

Image 147 Above: LZ 85 overhead the Zeppelin airship hangar at Neubeschenowa (see WIKIPEDIA Dudenstii Noi)

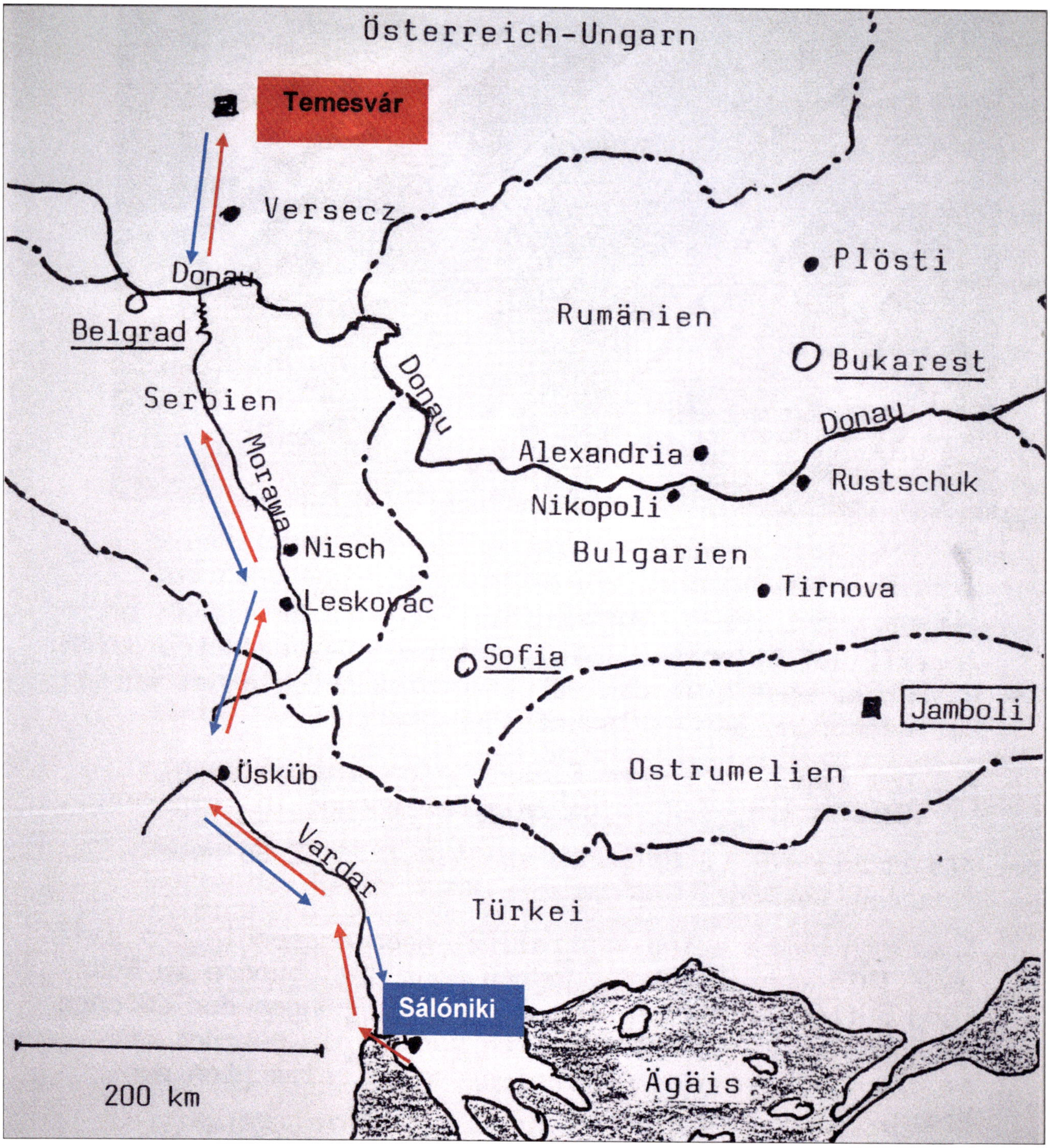

Image 1488 Above: LZ 85 attack on Sálóniki; January 31, 1916/February 1, 1916. (using a map from Navy, 1990, I, page 7)

due to a dense layer of clouds. As the loss of ground contact occurred at the same time as the airship was crossing overhead the ridge between Morawa and Vardar valley, a climb to the maximum altitude of 3000m was intiated. At 2350 the cloud layer gave way to clear conditions overhead the Vardar river as the airship continued its trip to Saloniki (SANKTANDRES, 1981, page 167).

The primary goal of this flight was to attack Saloniki harbor **(see image 149).**

"Heading south-southeast, the airship continued south of the city in order to get better acquainted with the locations of the harbor, the ships, etc. In addition to many smaller ships which faded from view there were several large, brightly lit vessels, probably transport category steamboats in the bay.

Image 149 Above: "Reconnaissance image of the Saloniki wharfage." (Mückler, 2013, page 106)

The airship turned toward the larger ships and the port facility housing the ammunition depot.

Multiple 60 kg bombs were dropped on the ships; one of the steamboats was hit on its port side...the largest number of bombs were dropped on the port facility and the railway near the harbor. Two bombs exploded on the pier head, six in the harbor. Multiple bombs hit the quays, causing massive explosions. One of the last bombs to hit the quays sparked a fire which spread quickly.

The airship was only sparsely shot at as the attacked appeared to take everyone on the ground by complete surprise. Proceeding on a northwesterly course, the airship reached the Vardar river around 0250, returning to Timisoara on the same route it followed on its way to Saloniki. As the southerly winds completely subsided and shifted north at times, the return was delayed by two hours, resulting in the implementation of further fuel preservation measures; part of the return trip was flown on only two engines to preserve fuel" (Navy, 1990, I, page 4).

Attacking Saloniki was of great importance as the Entente had landed there, trying to establish a Naval base for French and English warships. The airships' high transport capacity enabled the crews of "LZ 85" to drop two tons of bombs on the Saloniki Navy base. Following a 20-hour operating time, "LZ 85" approached Timisoara airship base (see ADZ, 4-2014).

On the morning of February 1, 1916, the return of "LZ 85" from its bomb attack flight on Saloniki was announced. Paul Biber made a note (see image 151) on the back of a photograph taken of the airship (see image 150), describing the circumstances of the day: "...bomb attack on Saloniki! The airship is visible on the horizon but already flying quite low of an altitude. Ground personnel running to meet the airship waded knee-deep in icy water, but a small "thump" was unavoidable! February 1, 1915 at approximately 1130am" (see images 150 through 153).

"The airship successfully landed at Szentendras at 1030am, gez. Scherzer..." the payload carried on board was as follows: 16-person crew (1300 kg), operating materials for 18 hours 3600 kg, water 6000 kg, bombs 2000 kg, MG armament 400 kg" (SANKTANDRES, 1981, pages 167/168).

The crew of "LZ 85" returned from their successful war patrol flight. Fortunately, Paul Biber made several notes regarding important occurrences on the back of his photographs **(see images 151 and 153).**

Image 150 Above: Airship "LZ 85" in range for arrival on February 1, 1916 während der Anfahrt zur Landung am 1.2.1916 (Foto: Archiv Biber).

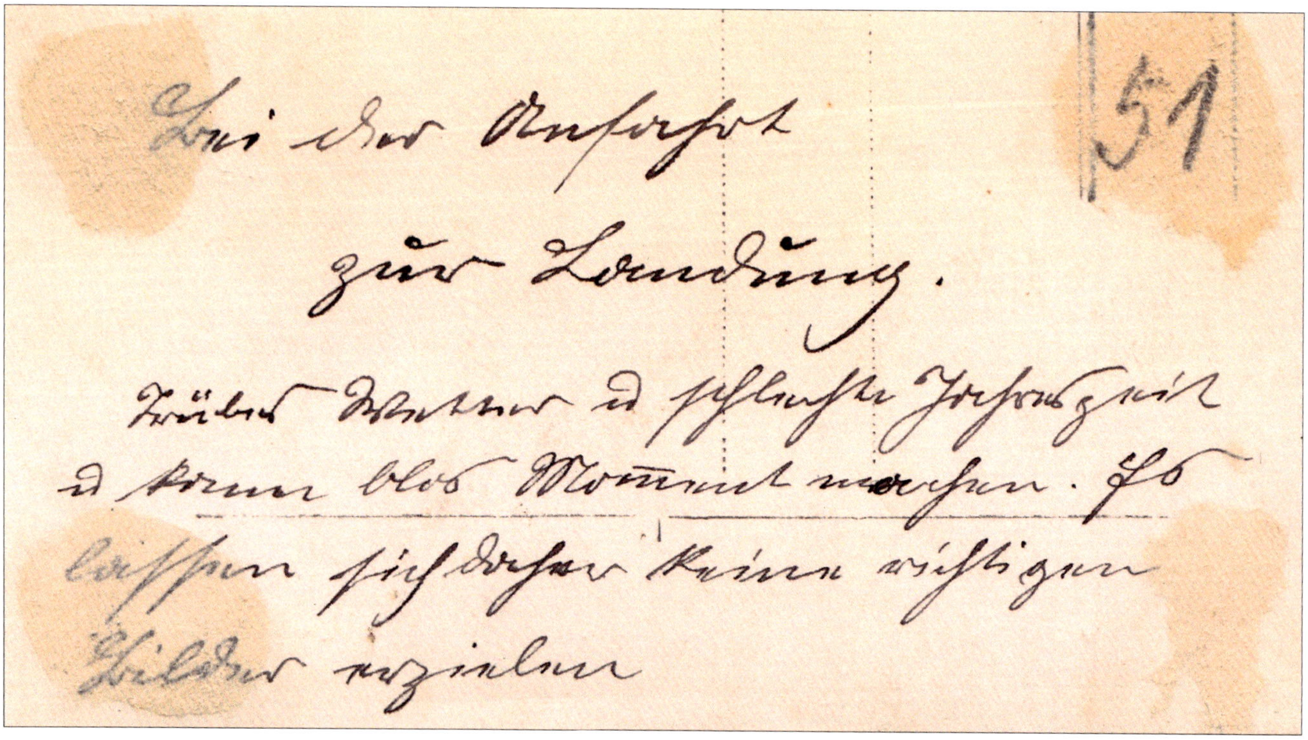

51

Bei der Anfahrt
zur Landung.
Trübes Wetter u. schlechte Jahreszeit
u. kann bloss Moment machen. So
lassen sich daher keine richtigen
Bilder erzielen

Image 150 Above: Reverse of photograph pictured in image 97, including the following note: "In range for arrival…" (Biber Archive).

Image 152 Above: Airship "LZ 85" approaching to land no February 1, 1916 at a low altitude (image: Biber Archive).

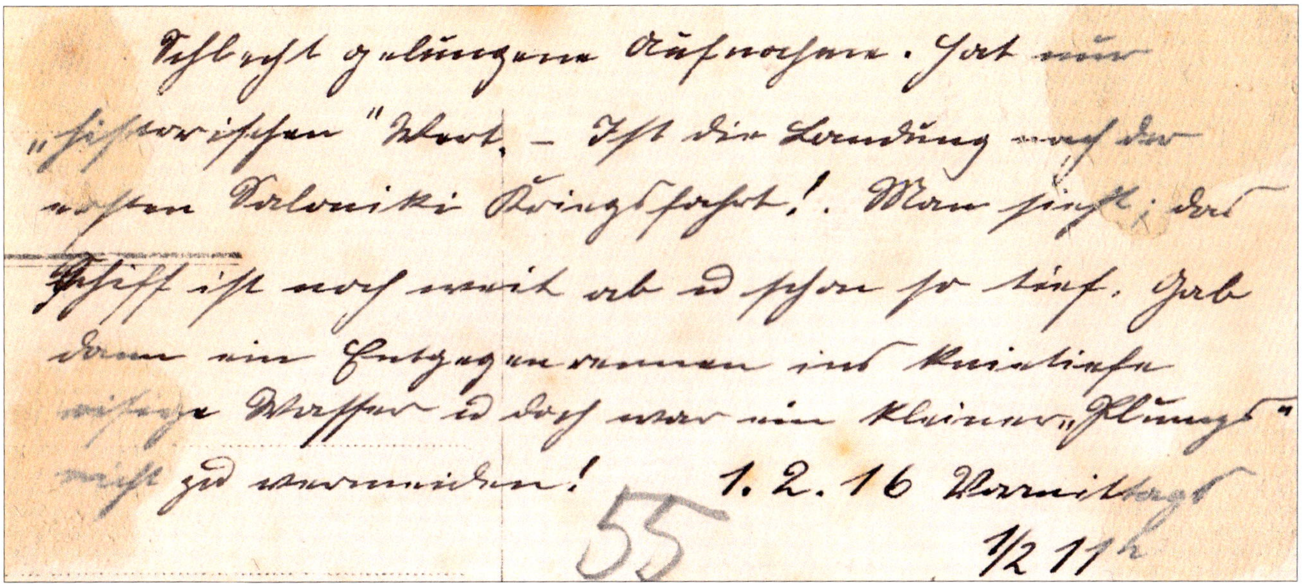

Image 153 Above: Reverse side of photograph pictured in image 99, including the following note: "...bomb attack on Saloniki! The airship is visible a great distance away but already at a low altitude. Ground crews ran to meet the airship, wading in knee-deep icy water were unable to avoid a small "thump" February 1, 1916, 11:30am." (Biber Archive).

After 18 hours in the air, "LZ 85" Kommandant Ernst Scherzer and his crew returned to base unscathed. In analyzing the successful bomb attack, the commander of "LZ 85" considered a fuel-stop in Sofia on the return flight from the next bomb attack on Saloniki. He would require less fuel for the shorter trip and could therefore carry additional bombs onboard "LZ 85".

5.2 Issues During "LZ 85" Bomb Attack on March 17, 1916

On February 8, 1915, "LZ 85" departed on its second flight to bomb Saloniki. Kommandant Scherzer was forced to turn the airship around and return to Timisoara due to poor weather conditions near Nisch. Paul Biber's notes on the back of photographs indicate that a few days after, orders were given to move "LZ 85 out of the hangar". After the airship had been carefully moved out of the hangar and brought into a favorable departure position along the tracks, everyone awaited the "Luftschiff hoch!" command. But soon an important message was passed along from meteorological service – "Storms approaching!" The subsequent command issued was "Luftschiff zur Halle. Marsch!"

It was not until the war patrol on March 17, 1916 that "LZ 85" and its crew flew to Saloniki again (see ADZ, 4-2014).

"The flight departed at 1825. By 1940, the airship had reached the Danube river; from there, the vessel turned toward the south, heading for the Vardar river. Navigation was easy due to the moonlight and clear weather conditions. The airship followed the Vardar valley and identified the town of Negotin before encountering a thick layer of clouds. As the mountain peaks were visible above the clouds and there were breaks in the clouds overhead the Vardar, the flight continued without any issues in hopes to locate Saloniki beneath the cloud deck. The bearing from mountain peaks and the Vardar for reference indicated that the airship was picking up speed at a surprisingly fast rate. The northerly winds were increasing. Strong northerly winds often picked up for several hours in the Vardar valley. When the Vardar no longer indicated any lift of the cloud deck, this was the indication that the airship was now near the river estuary. The airship was traveling at an altitude of 2200 m as the overcast cloud layer below them provided reliable protection from views from the surface.

At 0115, the airship was suddenly under strong fire associated with spotlight operation. As the crew of "LZ 85" was unable to detect a more suitable target and did not anticipate locating one with certainty, bombs were dropped in response to the gun shots from the ground.

As the gunfire ceased, the airship resumed its southeasterly course. After flying for 20 minutes, the airship was assumed to be south of Kara Burun, according to bearings taken from mountain peaks (Mount Olympus, Ossa, and Chalcidice). At 0145, orders were issued to turn toward a north-northwesterly course. Shortly after, the airship was

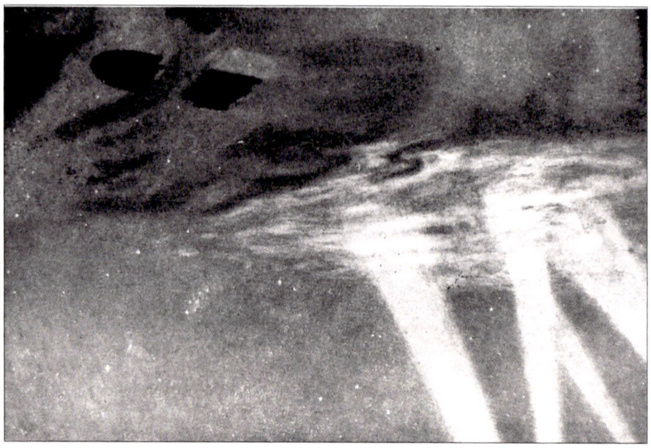

Image 154-1 Above: "LZ 85" is captured by spotlights in spite of the cloud cover (image: Schmeelke Archive).

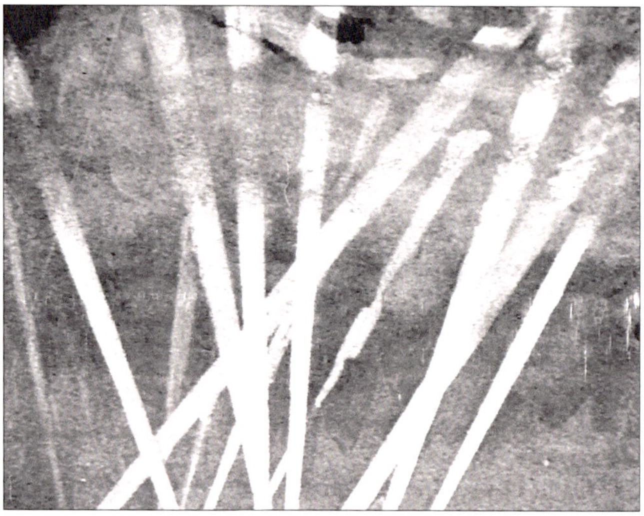

Image 154-2 Above: Many spotlights illuminate the sky in search of "LZ 85" (image: Schmeelke Archive).

searched for by 10-15 spotlights which located the vessel in spite of the cloud cover. (see images 154-1 and 154-2) It appears the airship had been spotted by batteries and ships due to its shadow above the clouds. The defensive fire was quite accurate in terms of direction and altitude.

The concentration of the gunfire and the spotlights left no doubt that LZ 85 was located above the fleet. The entirety of the remaining droppable ammunition (1200 kg) was dropped in the area where gunfire and spotlights were more detectable. At 0220 the bomb drop was concluded. Confirmation of hits could not be made through the cloud deck." (Navy, 1990, I, pages 5/6).

During the most recent bomb attack on Saloniki, "LZ 85" was met by a rather well-organized English and French air defense; a term which was coined during WWI. As a result of the focused air defense supported by powerful spotlights, multiple hydrogen cells onboard the rigid airship were destroyed (see ADZ, 4-2014).

Image 154-3 Above: "A train car filled with airship engines", (see above and left). A train arrived and transported field mail home (Paul Biber is pictured on the far left). (image: Biber Archive)

The return to Timisoara proved to be more challenging than anticipated as the airship had been hit and was losing hydrogen. As a result, a divert to Sofia was initiated due to its close proximity and the route to this destination did not require crossing a mountain ridge (see Ploes, 2018, page 18).

"LZ 85" slowly turned further to the northwest. "At 0235, the starboard engine failed due to a breakage of the drive shaft. As the moon was covered by clouds it was difficult to make out any ground features. The town of Kamanova was identified around 0520. As a result of the starboard engine failure and intermittent shutdowns of the remaining engines, enough fuel had been preserved that a landing at Sofia would still have been possible at 11am in moderate northerly winds.

By 10am, the airship had reached Sofia. The city as well as the entire country of Bulgaria lie beneath a thick overcast cloud layer; fortunately, the geographic location of the landing field could be determined without a doubt based on the mountain peaks that were visible above the cloud deck. The airship landed at 1020." (Navy, 1990, I, page 6).

Bulgarian troops assisted with the landing of "LZ 85" in Sofia. After offsetting the hydrogen loss by emptying approximately 1000 supplied gas cylinders, Kommandant Scherzer had anticipated departing Sofia that afternoon for Timisoara. He realized, however, that based on the significant temperature drop the airship would not be able to depart using power from all engines while carrying the entire crew. As a result, the departure was postponed.

In order to safeguard the continuation of the trip, Kommandant Scherzer ordered his crew to remove the starboard engine of the aft nacelle, including the propeller and parts of the radio station, along with any other non-essential parts and equipment (see Navy, 1992, III, page 11 and 1992, I, page 6).

After repeated verification, Kommandant Scherzer confirmed at 2300 that the lift generated would be sufficient to begin the journey back to Timisoara with an 11-man crew. "At 0600, we reached the Danube river and by 0800, we had Timisoara in sight.

While on approach for landing, the well-balanced airship suddenly encountered heavy rain squalls; for a short time, the squalls strained the airship to the extent that the landing had to be carried out several hundred meters away from the landing Trupp as it could only be controlled with difficulty. The airship slowly drifted away but the Trupp quickly arrived at the scene to provide assistance. Soon, the airship was back under control. The crew had been on duty for 41 hours without any replacement." (see Navy, 1990, I, page 6).

Image 155 Above: Airship "LZ 85" during a trial run (image: see WIKIPEDIA Dudestii Noi).

Image 156 Above: Timisoara airship base (image: Racil Collection).

5.3 March–May 1916 – Extensive Repairs on "LZ 85", Followed by a Successful Trial Run

At the request of Kommandant Scherzer, the removed components (for example the starboard engine) which had been left behind returned to Timisoara by railway along with part of his crew. Soon, the train carrying the remaining crew members, engines, and spare materials arrived at Timisoara along with some mail. After unloading the train car, it was filled with field mail to be sent home (see image 154-3).

Timisoara airship base ground personnel was busy carrying out repairs on "LZ 85" in order to get it back into an airworthy condition. Completing required tasks took up the entire month of April. In order to complete the necessary repairs, hydrogen was released from the damaged cells of the airship. The damaged cells were replaced and repaired, the engines reattached to their nacelles, and test runs were carried out. Finally, the cells were replenished with hydrogen.

It was not until the beginning of May that "LZ 85" was ready to be moved out of the hangar again. Initial ascent and test runs (see image 155) were successful. The airship was reported to be operational to Kommandant Scherzer (see Ploes, 2018, page 18). Aerial view images were taken of the occasion (see image 156).

Airship "LZ 85" officers such as Kommandant Oberleutnant Scherzer, Leutnant Serb, Oberleutnant Weidling, Hauptmann Kühn, Oberleutnant Nippe, Oberstleutnant Graf von Breda, Leutnant Adler, Leutnant Noak, and Oberleutnant Czernil (see image 157), expressed to Königlich Sächsischer Feldtrupp für Luftschiffe Nr. 14 – "LT 14" their gratitude for a job well done, particularly to the members of Schiffpflegetrupp (see Navy, 1992, I, page 10).

Hauptmann Kornel Poppe, Kommandant of ADEBAR himself participated in one of the "LZ 85" war patrol flights on Saloniki (see ADZ 4-2014 and Kustan & Penz, page 18).

5.4 Construction of an Outdoor Area Surrounding the Airship Hangar Including its Inauguration Ceremony

Once the soldiers had completed their daily duties of safeguarding air operations, they had the opportunity to discover the surrounding area of the airship base (see images 158 through 160). On February 27, 1916

Image 157 Above: Szentandrás airship base, 1916. Pictured from left to right; Leutnant Serb, Oberleutnant Weidling, Hauptmann Kühn, Oberleutnant Nippe, Oberstleutnant Graf von Breda, Kommandant Oberleutnant Scherzer, Leutnant Adler, Leutnant Noak, and Oberleutnant Czernil. Image part of the estate of FML Eugen Satori von Villemontain via Ing. Stefan F. Kozelka, Vienna (image: see Kustan & Penz, page 17).

Paul Biber made the following note on the left side of a photograph (see image 158): "Our quarters are located on the lonely puszta (Hungarian steppe). The town of Zent-Andrasch is on the right edge, made up of the small white dots. The clouds turned out well."

Airship base leadership was highly concerned with the appearance of the area surrounding the hangar, as well as to promote possibilities for ground crew self-sufficiency. Therefore, individual crews were encouraged to establish vegetable patches. Furthermore, crews created flower beds and built small garden sheds (see images 161 and 162).

Airship base Timisoara ground crew established nine different garden areas, each with a different theme.

On the occasion of the horticulturally designed outdoor area in front of the hangar, the "LT 14" crew gathered for a celebration (see images 163 through 167).

Timisoara airship base commander, Hauptmann Kornél Poppe and other leadership officers (see image 167a and 167b) were impressed by the beautiful presentation of the outdoor area. When looking at photographs of the horticulturally designed outdoor areas surrounding the airship hangar, it was hard to believe that these images were taken during a time of war. The images made a rather idyllic impression of the airship base. The entire "LT 14" crew was delighted by the blooming outdoor area.

Trupp "LT 14" soldiers also established a new tower to house a directional beacon at a greater distance from the airship hangar (see image 168). Russians aided "LT 14" ground personnel in the extensive works carried out to maintain the paths and roads of the airship base installation, as well as to care for the outdoor areas and other daily tasks (see images 169 through 172). The reverse of the photograph of the harbor, Paul Biber made the following note: "Stables and car shelter are pictured in the background- Kamerad Ascher (now part of the weather service) and a Russian." (see image 172).

Image 158 Above: Airship hangar surrounded by watery marsh, February 27, 1916; note on the left side (image: Biber Archive).

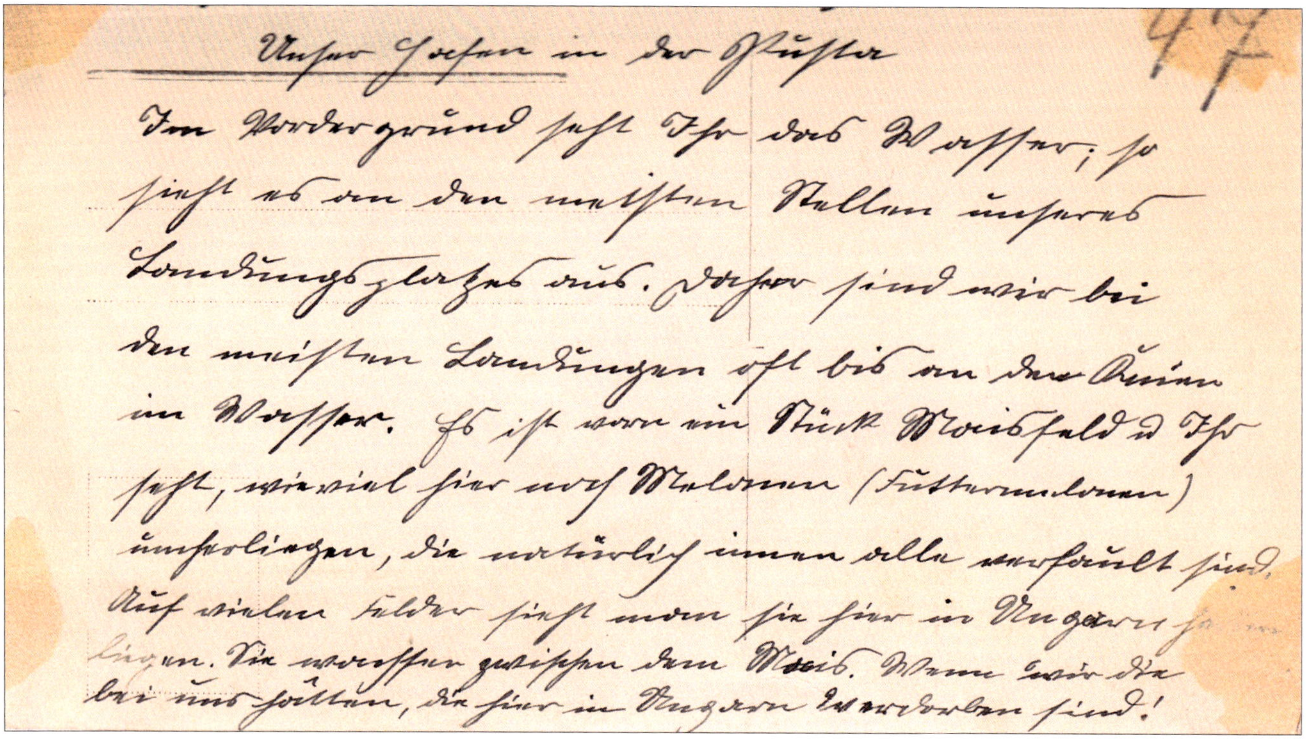

47

Unser Hafen in der Pußta

Im Vordergrund seht Ihr das Wasser; so
sieht es an den meisten Stellen unseres
Landungsplatzes aus. Daher sind wir bei
den meisten Landungen oft bis an die Knie
im Wasser. Es ist war ein Stück Maisfeld u. Ihr
seht, wieviel hier noch Melonen (Futtermelonen)
umherliegen, die natürlich immer alle verfault sind.
Auf vielen Felder sieht man sie hier in Ungarn so
liegen. Die wachsen zwischen dem Mais. Wenn wir die
bei uns hätten, die hier in Ungarn verdorben sind!

Image 159 Above: Reverse of photograph in image 158 (Biber Archive).

„Aufgenommen am Sonntag, d. 27.2.16
Unser Jochen in der Pusta
Im Vordergrund seht Ihr das Wasser, so sieht es an den meisten Stellen unseres Landungsplatzes aus. Daher sind wir bei den meisten Landungen oft bis an den Knien im Wasser. Es ist vorn ein Stück Maisfeld und Ihr seht, wieviel hier noch Melonen (Futtermelonen) umherliegen, die natürlich innen alle verfault sind. Auf vielen Feldern sieht man sie hier in Ungarn jetzt liegen. Sie wachsen zwischen dem Mais. Wenn wir die bei uns hätten, die hier in Ungarn verdorben sind!"

Image 160 Above: Text on the reverse of Image 159.

Image 161 Above: Comrades doing yardwork outside of the airship hangar (image: Biber Archive).

5.5 May 1916 – "LZ 85" Last War Patrol Flight

Following the completion of extensive and lengthy repairs carried out by Luftschifftrupp "LT 14", "LZ 85" was scheduled for new orders to drop bombs on Saloniki on May 4, 1916. Timisioara airship hangar was bustling with activity all day. Railway cars had delivered bombs which needed to be offloaded (see image 173-1), transported to the airship and attached inside the vessel. (see image 173-2)

Luftschifftrupp was busy preparing "LZ 85" for

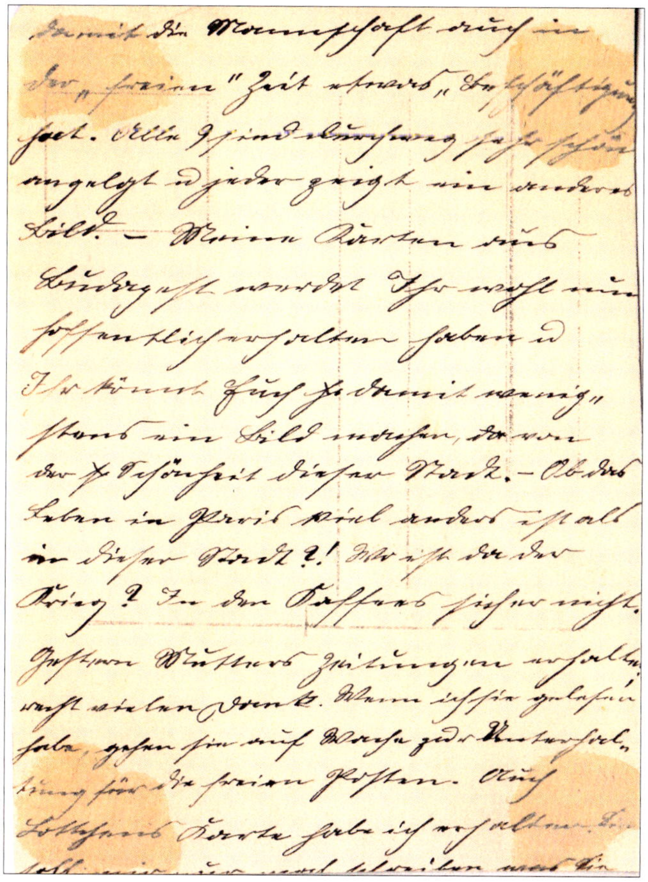

damit die Mannschaft auch in
der „freien" Zeit etwas „Beschäftigung"
hat. Alle sind durchweg sehr schön
angelegt u. jede zeigt ein anderes
Bild. – Meine Karte aus
Budapest werdet Ihr wohl nun
hoffentlich erhalten haben u.
Ihr könnt Euch so damit wenig=
stens ein Bild machen, der von
der Schönheit dieser Stadt. – Ob das
Leben in Paris viel anders ist als
in dieser Stadt?! Wo ist da der
Krieg? In den Kaffees sicher nicht.
Gestern Mutters Zeitungen erhalten,
recht vielen Dank. Wenn ich sie gelesen
habe, gehen sie auf Wache zur Unterhal=
tung für die freien Posten. Auch
Lottechens Karte habe ich erhalten.
Soll mir nur mal schreiben was sie

„Damit die Mannschaft auch in der „freien" Zeit etwas „Beschäftigung" hat. Alle sind durchweg sehr schön angelegt u. jede zeigt ein anderes Bild. Meine Karte aus Budapest werdet Ihr wohl nun hoffentlich erhalten haben u. Ihr könnt Euch damit wenigstens ein Bild machen von der Schönheit dieser Stadt. Ob das Leben in Paris viel anders ist als in dieser Stadt?! Wo ist da der Krieg?! In den Kaffees sicher nicht. Gestern Mutters Zeitungen erhalten, recht vielen Dank. Wenn ich sie gelesen habe gehen sie auf Wache zur Unterhaltung für die freien Posten. Auch Lottechens Karte habe ich erhalten. Soll mir nur mal schreiben, was sie …

Image 162 Facing Page Above: Comrades established plant patches for "LT 14" (image: Biber Archive).

Image 163 Facing Page Below: Comrades gathered for a celebration on the occasion of the inauguration of the outdoor area surrounding the airship hangar (image: Biber Archive).

Image 164 Top: Group photograph of the entire "LT 14" crew taking possession of the outdoor area (image: Biber Archive).

Image 165 Left: Reverse of photograph taken by Paul Biber in image 164 (Biber Archive).

Image 166 Above: Text in image165.

this important war patrol flight to the best of their abilities. After expeditiously moving the airship out of the hangar, Luftschifftrupp 14 maneuvered "LZ 85" into the desired position for departure.

The LT 14 ground crew still held on to the "LZ 85", keeping it in balance and on the ground (see image 174). Handrails were attached to the nacelle that ground crews grabbed hold of in order to securely keep the airship in balance. Everyone awaited the start command.

At 1630, the "hoch!" order was issued as the fully loaded airship departed toward the Morawa valley. (see Ploes, 2018, page 18)

This time the airship was to attack Saloniki from over the ocean as it was assumed that its ground defense was anticipating another attack from the mainland. Defensive actions overall had been expanded and coordinated in Saloniki.

At 0215 am, "LZ 85" arrived overhead the Saloniki harbor. Alarm was given immediately; in response to the airship's attack from overhead the ocean, a focused defense was initiated using searchlights and air defense. "LZ 85" was caught in a crossfire. In addition to anti-aircraft artillery, war

Image 167a Facing Page Above: Hauptmann Kornél Popp (far left,in front) with other officers by the gardens (image: Biber Archive).

Image 167b Above: Commanding officers from Timisoara airship base presented themselves at the inauguration of the gardens which "LT 14" members had created adjacent to the airship hangar, spring 1916. Hauptmann Kornel Popp, Timisoara airship base Commander, is pictured in the front row, second from right. (image: Biber Archive).

Image 168 Facing Page Below: Construction of a new tower to house the directional beacon (image: Biber Archive).

Image 169 Above: The "Russians" and "us" (image: Biber Archive).

Image 170 Below: Russians working on the airship base (image: Biber Archive).

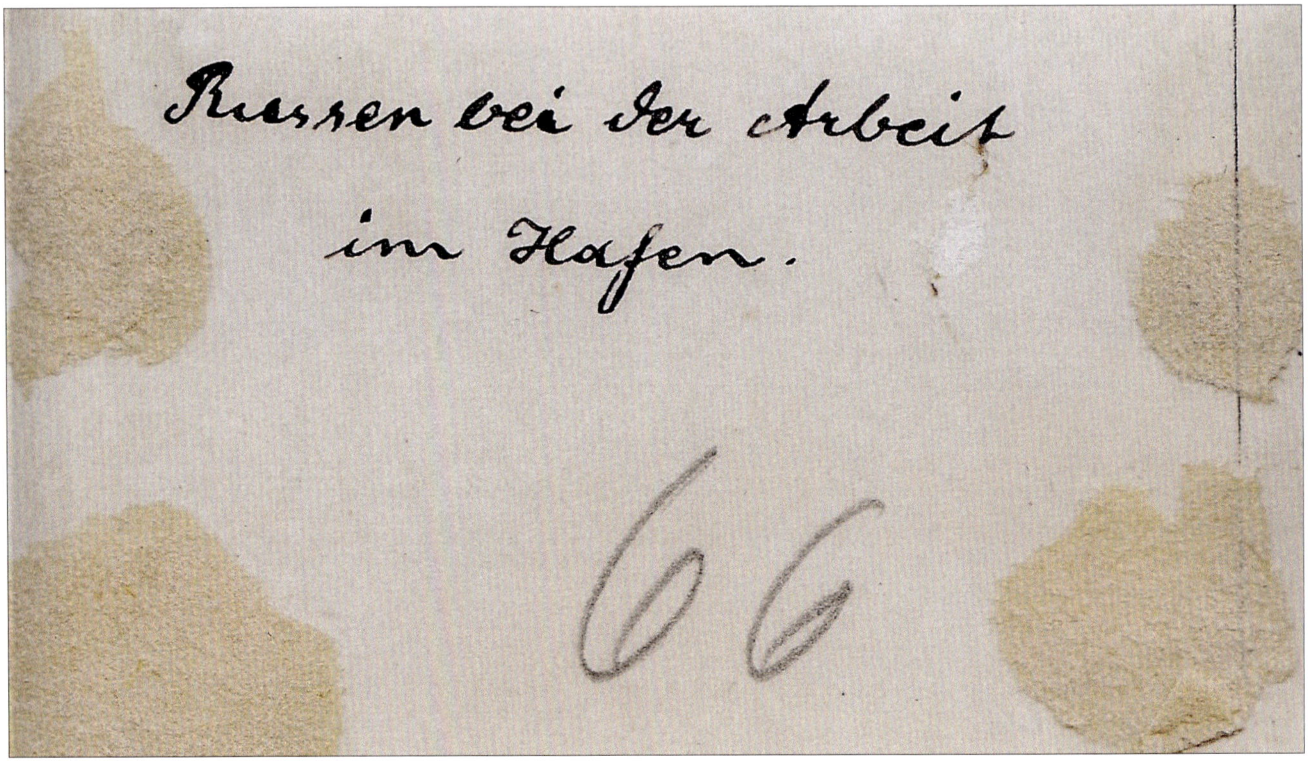

ships were on the alert in the harbor of Saloniki. "LZ 85" was hit by MMS Agamemnon. Slowly descending, the airship turned toward the south, "where the English torpedo boat H.B.T.B. 18 gave "LZ 85" its mortal blow by means of three precise shots." (see Ploes, 2018, page 19)

"LZ 85" was slowly crashing. The crew attempted to maneuver the airship toward the southwest in order to reach Greece. This endeavor put excessive strain on the forward engines that were still operable. All removable, accessible equipment such as airship materials, machine guns, dropping devices, radio communication devices, etc. were thrown overboard, to no avail. At 0245, LZ 85 touched down hard at the edge of the Vardar river valley – in the reeds of the Vardar estuary (the Vardar-swamps) – approximately 25 km from Saloniki. The rear of the airship was lie deep in the swamp and had been filled with water; the rear nacelle was no longer accessible. The tip of LZ 85 protruded about 15 m into the air. These circumstances did not allow for the sawing off of the rear nacelle in an attempt to continue the trip. Though frightened, all 12 crew members were unharmed. Confidential materials were destroyed. Hauptmann Scherzer and Oberleutnant Nippe were the last crew members to leave airship LZ 85. From a distance of approximately 30 meters, Hauptmann Scherzer shot the airship ablaze. The blast of the resulting explosion threw Hauptmann Scherzer

Image 173-2 Above: Attaching bombs on the outside of Zeppelin "LZ 85" (image: Schmeelke Archive).

Image 171 Facing Page Above: Reverse of image 170 – "Russians working at the harbor." (Biber Archive)

Image 172 Facing Page Below: Image on the harbor with a note on the reverse side (Biber Archive).

Image 173-1 Above: Offloading ammunition and bombs (image: see scharekf.hu)

Image 174 Below: The Haltemannschaft (holding crew) near the nacelle of LZ 85 (image: see scharekf.hu)

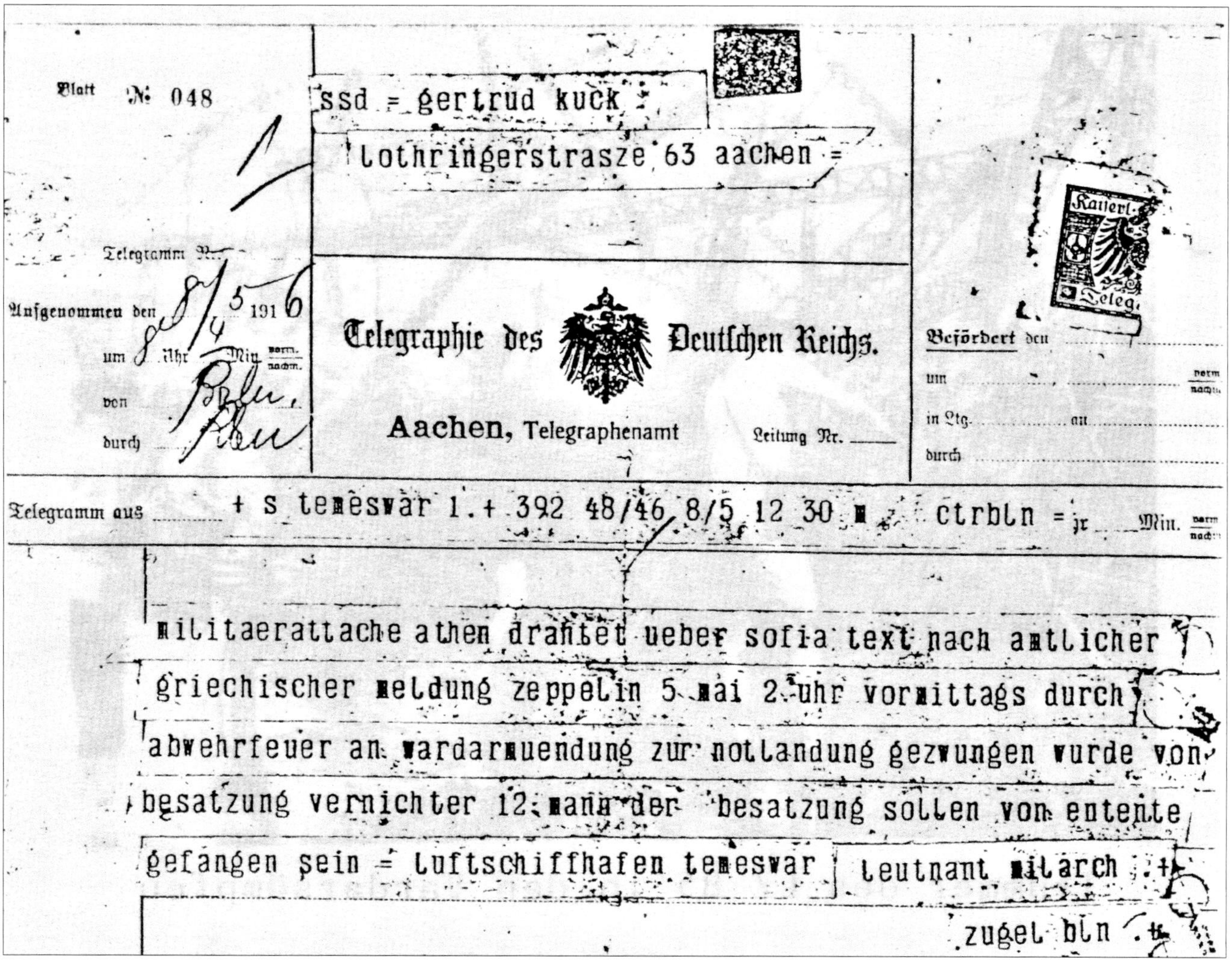

Blatt № 048

ssd = gertrud kuck =
lothringerstrasze 63 aachen =

Telegramm Nr.
Aufgenommen den 8/5 1916
um 8 Uhr Min. vorm. nachm.
von Berlin
durch Blu

Telegraphie des Deutschen Reichs.
Aachen, Telegraphenamt
Leitung Nr.

Befördert den
um vorm. nachm.
in Ltg. an
durch

Telegramm aus + s temeswar 1 + 392 48/46 8/5 12 30 m ctrbln = jr Min. vorm. nachm.

militaerattache athen drahtet ueber sofia text nach amtlicher griechischer meldung zeppelin 5 mai 2 uhr vormittags durch abwehrfeuer an wardarmuendung zur notlandung gezwungen wurde von besatzung vernichter 12 mann der besatzung sollen von entente gefangen sein = luftschiffhafen temeswar leutnant milarch .+

zugel bln .+

Image 175 Facing Page Above: "LZ 85" wreckage (image: Navy, 1992, I, page 3).

Image 176 Facing Page Below: Posing with the naval ensign of "LZ 85" (image: see scharekf.hu)

Image 177 Above: Notification of "LZ 85" family members (Navy, 1992/I, page 4)

Image 178 Right: "LZ 85" crew members while in captivity in Saloniki. Hauptmann Scherzer, commander of airship "LZ 85" (middle, 5th from left) (image: Mückler, 2013, page 107)

Wreckage of "LZ 85" in front of Saloniki's white tower (image: Navy, 1992/I, page 5)

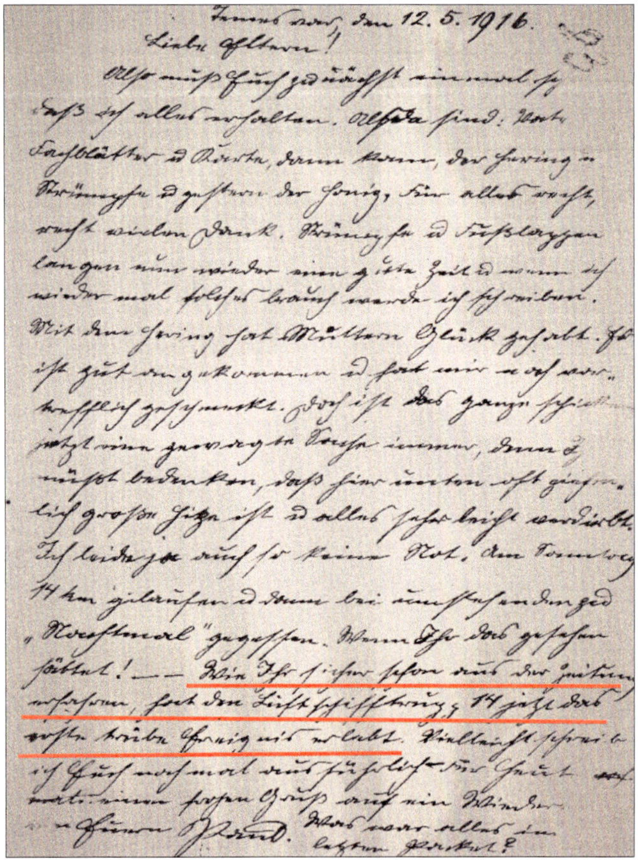

Temeswar, den 12.5.1916

Liebe Eltern!

Also muß ich Euch zunächst einmal bestätigen, dass ich alles erhalten. Als da sind Vaters Fachblätter und Karten, dann kam der Hering und Strümpfe und gestern der Honig, für alles recht, recht vielen Dank. Strümpfe und Fußlappen langen nun wieder eine gute Zeit und wenn ich wieder mal solche brauche, werde ich schreiben. Mit dem Hering hat Muttern Glück gehabt. Er ist gut angekommen und hat mir schon vortrefflich geschmeckt. Doch ist das ganze Schicken jetzt eine gewagte Sache immer, denn Ihr müßt bedenken, daß hier unten oft ganz viel große Hitze ist und alles sehr leicht verdirbt. Ich leide ja auch so keine Not. Am Sonntag 14 km gelaufen und dann bei Umstehenden zu „Nachtmahl" gegessen. Wenn Ihr das gesehen hättet! – Wie Ihr sicher schon aus der Zeitung erfahren hat der Luftschifftrupp 14 jetzt das erste trübe Ereignis erlebt. Vielleicht schreibe ich Euch nochmal ausführlich.

Einen frohen Gruß auf ein Wiedersehen

Von Eurem Paul

Was war alles im letzten Paket?

Image 180 Left: Field postcard from Paul Biber – Timisoara, May 12, 1916. Reverse of image 192 (Biber Archive)

Image 181 Above: Text from image 180.

and Oberleutnant Nippe backwards. The metal frame of "LZ 85" was the only thing remaining **(see image 175).** The crew were moving in a south-southwesterly direction, wading through waist deep swamplands or thistle fields, trying to reach Greek territory. The fleeing crew members were spotted by an aircraft in a crop field at 0930 am. Soon after, the entire crew were surrounded by French and Serbian cavalry and taken captive on May 5, 1916. The story made the headlines of the world press (see ADZ, 5-2014 and Navy, 1992, I, pages 2 through 4). They were particularly proud to have gotten their hands on the intact "LZ 85" naval ensign. Everyone wanted to pose for pictures with the naval ensign. (see image 176)

Family members of the LZ 85 crew were notified of the imprisonment via telegram from Timisoara (see image 177).

"The French war journal *'La Guerre Aérienne illustrée'* proudly reported of the imprisonment of crew members whose German war airship "LZ 85" had crashed in the Vardar swamp after being shot down by the English battleship 'Agamemnon' on May 5, 1916" (Mückler, 2013, page 107).

Crew members were held captive in Saloniki where they were interrogated (see image 178). Later, the prisoners were distributed to 5 different camps throughout large areas of Southern France.

The prisoners were questioned by two English interrogation officers in France between December 15–20, 1916 (see Navy, 1992, I, pages 5 through 11).

The wreckage of LZ 85 was disassembled on-site in the swamps and brought to Saloniki with rafts and cargo barges in order to be displayed in front of the white tower. The frame of the airship formed an impressive backdrop for the curious population of Saloniki and the surrounding areas as well as for soldiers of different nationalities stationed in the area (see image 179) (see Ploes, 2018, page 21).

The loss of "LZ 85" and its crew under the command of the experienced Hauptmann Ernst Scherzer was a serious setback for combat readiness, resulting in an operational standstill of Szentandrás airship base. It also proved to be a painful experience for Luftschifftrupp 14 (see underlined text taken from field post excerpts dated May 12, 1916, in images 180 and 181).

According to Paul Biber's narrative on a field post card (see images 180 and 181), he hiked to a village on Sunday (May 12, 1916) and joined the local farmers for dinner. He appeared to enjoy his meal.

The non-operational state of Timisoara airship base needed to be changed immediately. Without delay, the replacement of "LZ 85" by "LZ 86" was initiated. "LZ 86" was withdrawn from the eastern front and its conversion (the airship's physical extension) was commissioned.

6. June Through Mid-August 1916 – Landing of "SL 10" and "LZ 101" at Timisoara Airship Base

6.1 June 1916 – Commissioning of Jamboli Airship Base in Bulgaria Using Airship "SL 10"

As a result of the loss of "LZ 85", the construction of a second German airship hangar in the Balkans (near Jamboli in Bulgaria, approximately 80 km west of Burgas) was expedited and rapidly completed in order to improve the exertion of influence on war events (see images 182 through 184). The airship base included a large hangar, rail connection, gas works, electrical power plant, and a radio station with a direction-finding installation. The small village of Jamboli was advantageously situated in the plains of the river Tundsha, with prevailing winds almost entirely facing the direction of the airship hangar. Summer temperatures of 40°C and higher proved to be challenging for the personnel (see Lehmann, 2019).

"Due to its advantageous location, airships were able to launch attacks on Russian harbors in the Black Sea and English footholds in the Aegean Sea. On the flight from Mannheim to Jamboli, "SL 10" Commander Hauptmann von Wobeser sent a telegraphic greeting to K.u.k. Navy and Air Force Command while the airship was passing overhead Vienna on June 21, 1916, at 08:00." (Kustan & Penz, page 18).

While there was no airship at the Szentandrás airship base for a period of time, everyone was aware that this was only a temporary situation. Airship troop "LT 14" carried out many tasks in order to maintain a high level of operational readiness at the base. The production of hydrogen continued. Hydrogen was then filled into gas bottles to be properly stored in the bunker; spare parts were serviced, engines dismantled in a way that would allow for their railway transportation to Germany where they were overhauled.

After completing their tasks, for example at the gas works, Luftschifftrupp "LT 14" members were able to take a refreshing dip in the nearby pool of the gas works (see image 185). The Artesian well belonging to the pool had been drilled to a depth of 200 meters.

Luftschifftrupp 14 members were overall content with the condition of their accommodations, workplaces (see image 186), their daily work routines, as well as their ability to remain in contact with their homeland (see images 187 through 190, and image 194) and the connections they formed with the local people (see images 191 through 193).

The train car transporting Luftschiffer field letters to the homeland also served as a means of transport for Leutnant Stigga's dog which had been held captive in Saloniki. According to a note by Paul Biber on the reverse of the field post card pictured in image 188, the dog was shipped to Luftschifftrupp dependents via railway.

On the reverse of his field postcard (see image 193), he made the following note: "Happy wedding anniversary to my parents! Love, Paul" (see image 194).

Word reached the airship base that airship "SL 10" would make a stopover at Szentandrás on June 21, 1916, while on its way to Jamboli airship base where it would be stationed. Luftschifftrupp "LT 14" carried out preparations for the impending layover of "SL 10".

On June 21, 1916, at 13:40, "SL 10" (see image 195-1) landed at Szentandrás airship base. It continued its journey to Jamboli on June 23, 1916, at 12:20 (see Kustan & Penz, page 18). The abbreviation "SL" stands for "Schütte-Lanz" airship. SL airships were rigid airships such as those built by Duke Ferdinand von Zeppelin. SL is short for the originators of this type of rigid airship construction, Professor Johann Schütte and Ingenieur Dr. Karl Lanz.

Contrary to Duke Ferdinand von Zeppelin's rigid airships, SL airships had a wooden frame whose construction in turn was based on a patent by Berlin resident Huber (see Griehl, 2010, page 43).

Schütte-Lanz-Luftschiff "SL 10" was part of the e construction project, with a gas content of 38,000 m^3. It began operations on May 17, 1916.

SL 10 was tasked with the reconnaissance of the Black Sea coastal areas departing from Jamboli airship base (Bulgaria), as well as the bombardment of British stations in the Greek Isles. Between late June and late July, the Schütte-Lanz airship carried out reconnaissance flights across the Black Sea. Its

Image 182-1 Above: Jamboli airship base (Bulgaria) as viewed from an airship (image: Schmeelke Archive).

Image 182-2 Below: Accommodation barracks on the grounds of Jamboli airship base; a radio mast is visible on the right-hand side of the image (image: Schmeelke Archive).

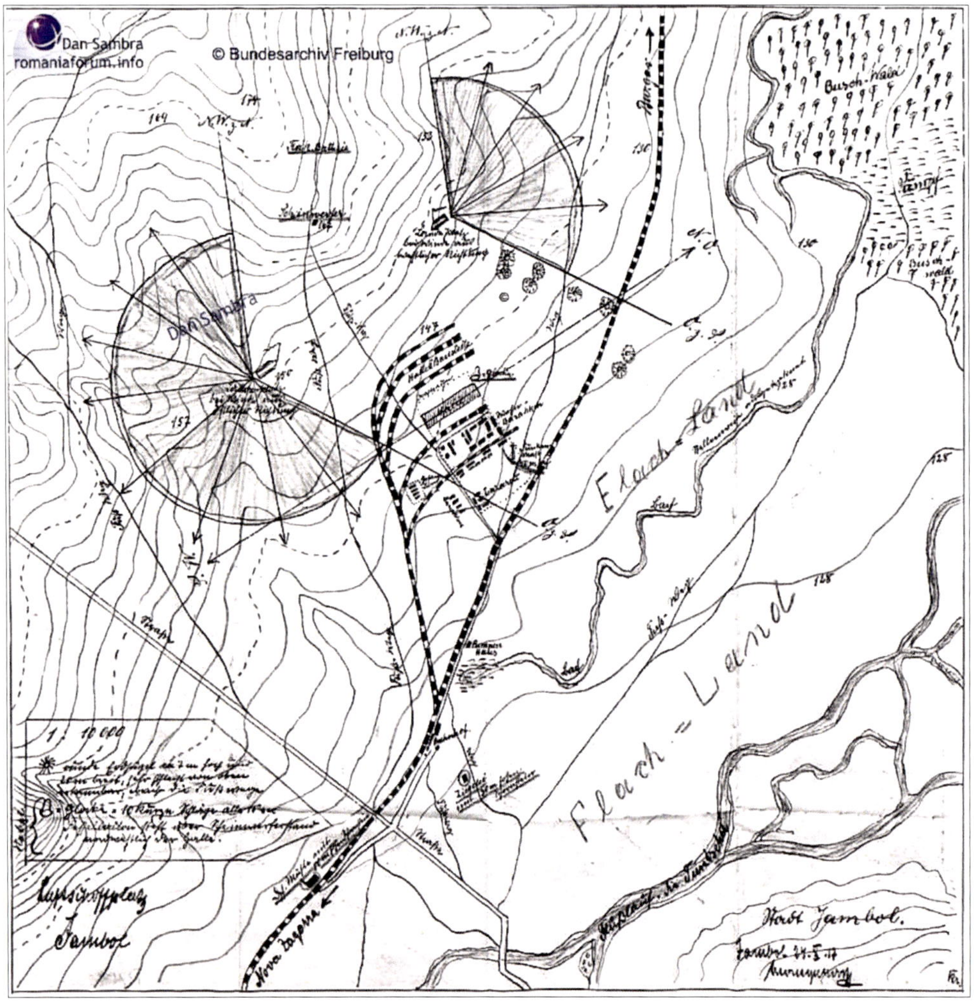

Image 183 Left: Map of Jamboli airship base (Bulgaria) – code name "BORIS" (Freiburg Federal Archives copy: Racil Collection)

Image 184 Below: Comrades in front of the harbor; note on reverse side (see image 164, underlined text): "Standing near the barracks and our machine building (where the crate is)" (image: Biber Archive)

Image 185 Above: Comrades enjoying a refreshing swim in the gas works pool (image: Biber Archive)

Image 186 Above: Jamboli airfield and airship base (Bulgaria) (image see scharekf.hu)

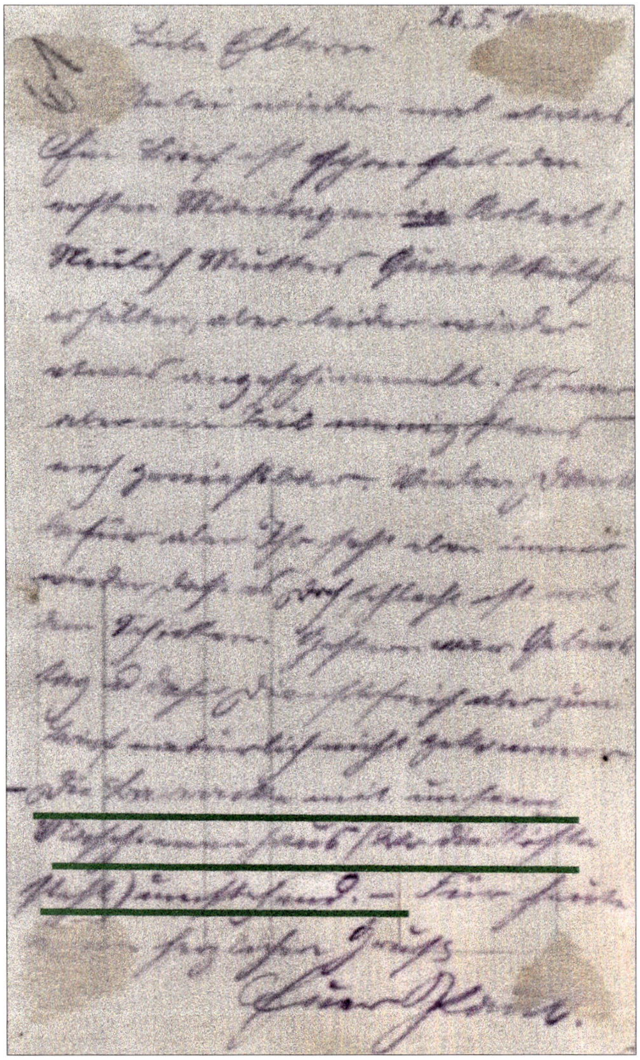

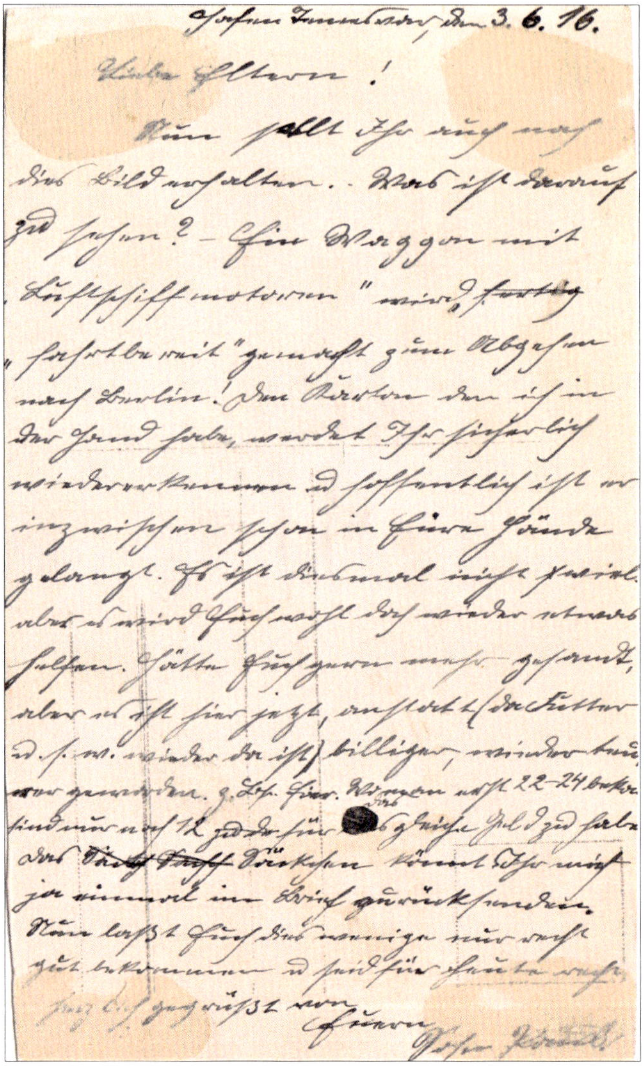

Hafen, Temesvár 3.6.16

„Liebe Eltern!

Nun sollt Ihr das Bild erhalten. Was ist darauf zu sehen? Ein Waggon mit „Luftschiffmotoren“ wird fahrbereit gemacht zum Abgehen nach Berlin. Den Karton, den ich in der Hand habe, werdet Ihr sicherlich wiedererkennen u. hoffentlich ist er inzwischen schon in Eure Hände gelangt. Es ist diesmal nicht viel, aber es wird Euch wohl doch wieder etwas helfen. Hätte Euch gern mehr gesandt, aber es ist hier jetzt, anstatt (da Futter u.s.w. wieder da ist) billiger, wieder teurer geworden. Wo man erst 22 - 24 bekam sind nur noch 12 für das gleiche Geld zu haben. Das Säckchen könnt Ihr mir ja einmal im Brief zurücksenden. Nun laßt Euch das Wenige nur recht gut bekommen u. seid für heute gegrüßt von

Eurem

Paul“

Image 188 Above: A train car headed for the homeland is loaded with mail written by aeronauts (Paul Biber is pictured on the left) (image: Biber Archive).

Image 187 Facing Page Above Left: Reverse side of photograph in image 186 (Biber Archive).

Image 191 Above: Group photo at a farm (image: Biber Archive).

Image 190 Facing Page Below: Text on reverse of image 189.

Image 189 Facing Page Above Right: Reverse side of photograph in image 188. (Biber Archive)

Image 192 Above: Picking up milk from the farm which was used for pancakes, Quarkkäulchen (a Saxon dish made from dough containing mashed potatoes, quark cheese, eggs, and flour with the optional addition of cinnamon or raisins), scrambled eggs, Eierschecke (a confectionery specialty from Saxony and Thuringia), etc. (image: Biber Archive).

Image 193 Above: Comrades Schlesinger and Krusemark collecting water at the "Puszta (Hungarian steppe) hut (Pustahäuschen) situated between the hangar and the village" (image: Biber Archive).

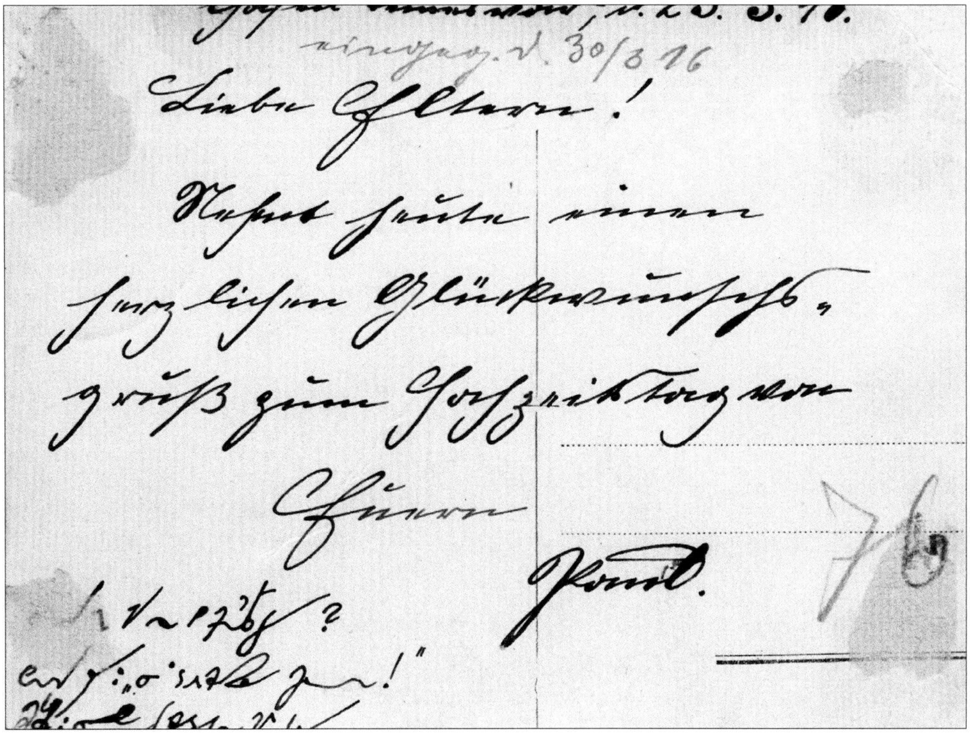

Liebe Eltern!

Image 194 Left: Reverse of photograph pictured in image 193, congratulations from Paul Biber to his parents (image: Biber Archive).

Image 195-1 Above: SL Assault airship (see Griehl, 2010, page 46).

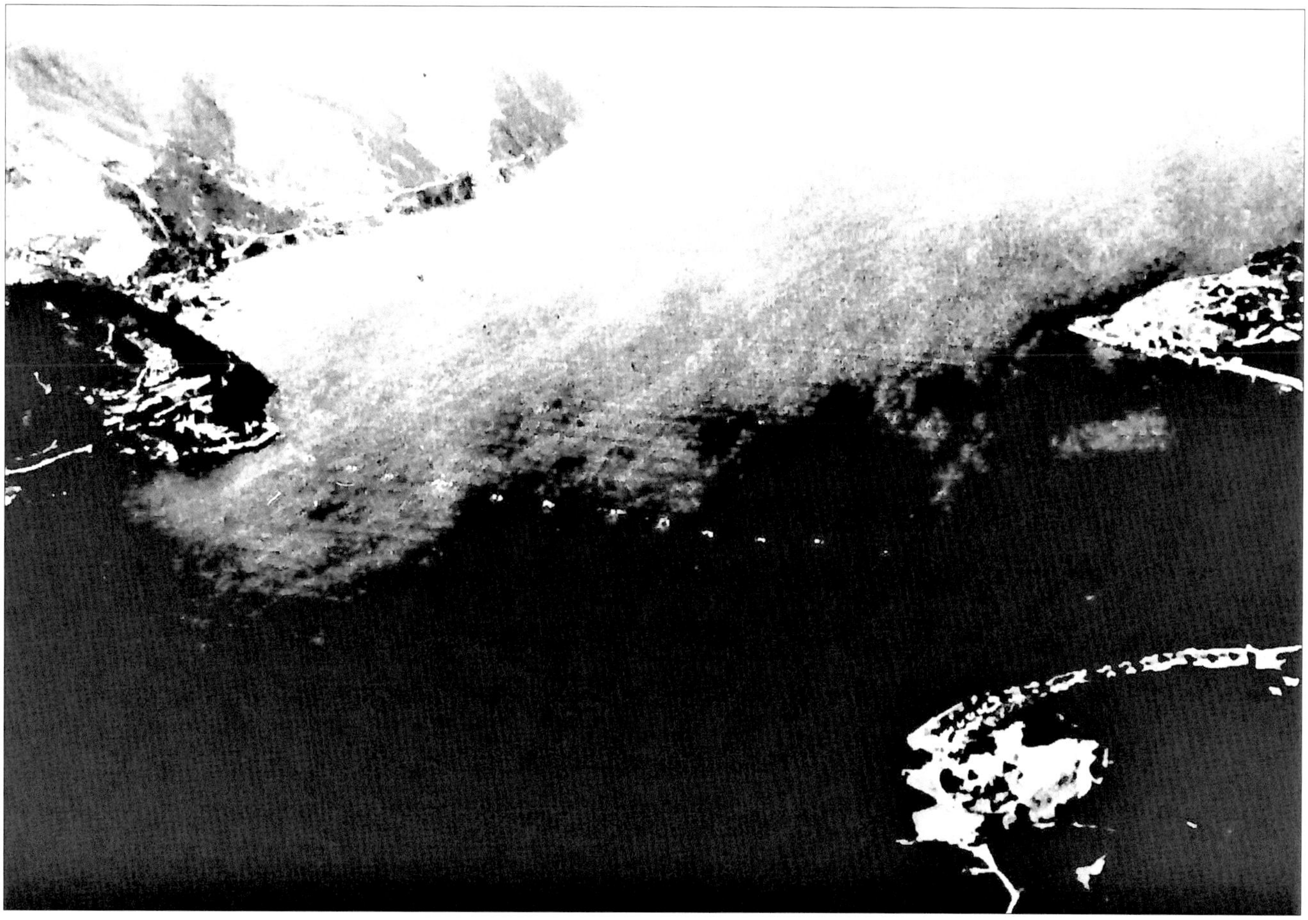

Image 195-2 Above: Submarine barrier on the Bosporus (image: Schmeelke Archive).

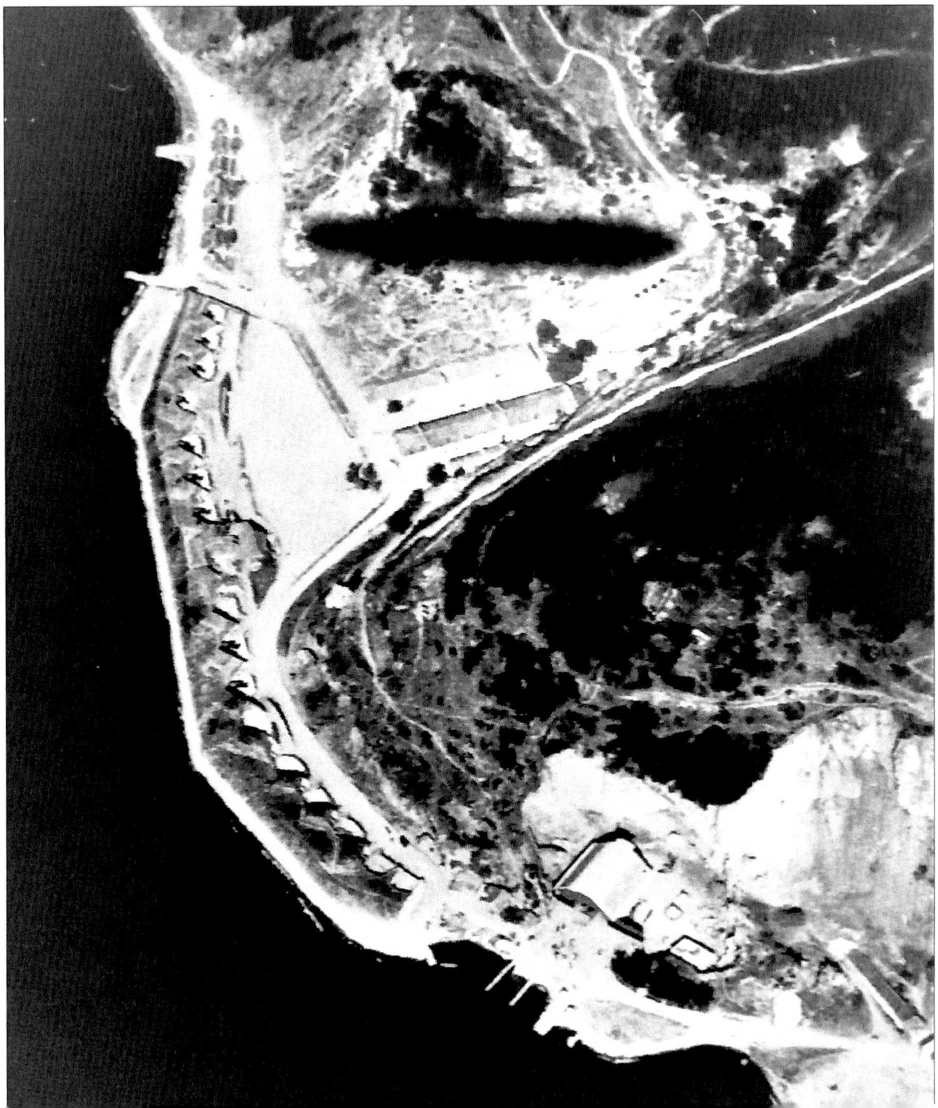

Image 195-3 Above Left: German warships on the Bosporus (image: Schmeelke Archive).

Image 195-4 Above Right: German warships in the port of Stenia on the Bosporus (image: Schmeelke Archive).

Image 195-5 Left: Bosporus fortification with a shadow cast by an airship (image: Schmeelke Archive).

Image 195-6 Above: Airship "SL 10" overhead Istanbul (image: Schmeelke Archive).
Image 196-1 Below: Field postcard written by Paul Biber, dated July 16, 1916. Reverse of photograph in image 191 (Biber Archive).

Image 196-2 Above: Postmark on Field postcard written by Paul Biber, dated July 16, 1916. (Biber Archive).

Image 197 Below: LZ 101 landing (image: Navy, 1989/IV, page 5).

Image 198-1 Facing Page Above: Paul Biber's military passport. (Biber Archive).

Image 198-2 Facing Page Middle: Paul Biber leaving Luftschifftrupp Nr. 14. (Biber Archive).

Image 199 Facing Page Below: Stamp of Paul Biber's new office (Biber Archive).

missions included searching for Russian minefields (see image 195-2) and explored the situation on the Bosporus (see images 195-2 through 195-5); on their journey, they also overflew Istanbul (see image 195-6). On July 28, 1916, it went missing over the Black Sea during an attempted attack on Sevastopol departing from Jamboli airship base (see Griehl, 2010, pages 46/47). It is assumed that the airship was hit by lightning during a storm, causing the wooden frame to catch on fire.

Among other notes, Paul Biber wrote the following on the back of a field postcard dated June 16, 1916 **(see image 123):**

"Above all, there is a bad atmosphere in the Truppe at the moment. While it may be a while until it gets better, change is not impossible at this point."

Paul Biber made the following note on a Timisoara field postcard dated July 16, 1916 **(see image 196-1 & 196-2,** framed in green):

"It appears as though my circumstances will be changing soon. I wanted to notify you of the correct address. Write only: Deutscher Feldtrupp f. Luftsch. Nr. 14. Feldpoststation 181. – If you add Temesvár, it might be sent via Hungarian mail which is not nearly as safe or fast. 181 is now our closest post office…"

While the designation of the new field postal code 181 **(see image 196-1 & 196-2)** was not an indication of a possible change of station for "LT 14", Paul Biber may have had a hunch that change was coming. Something new was on the horizon for

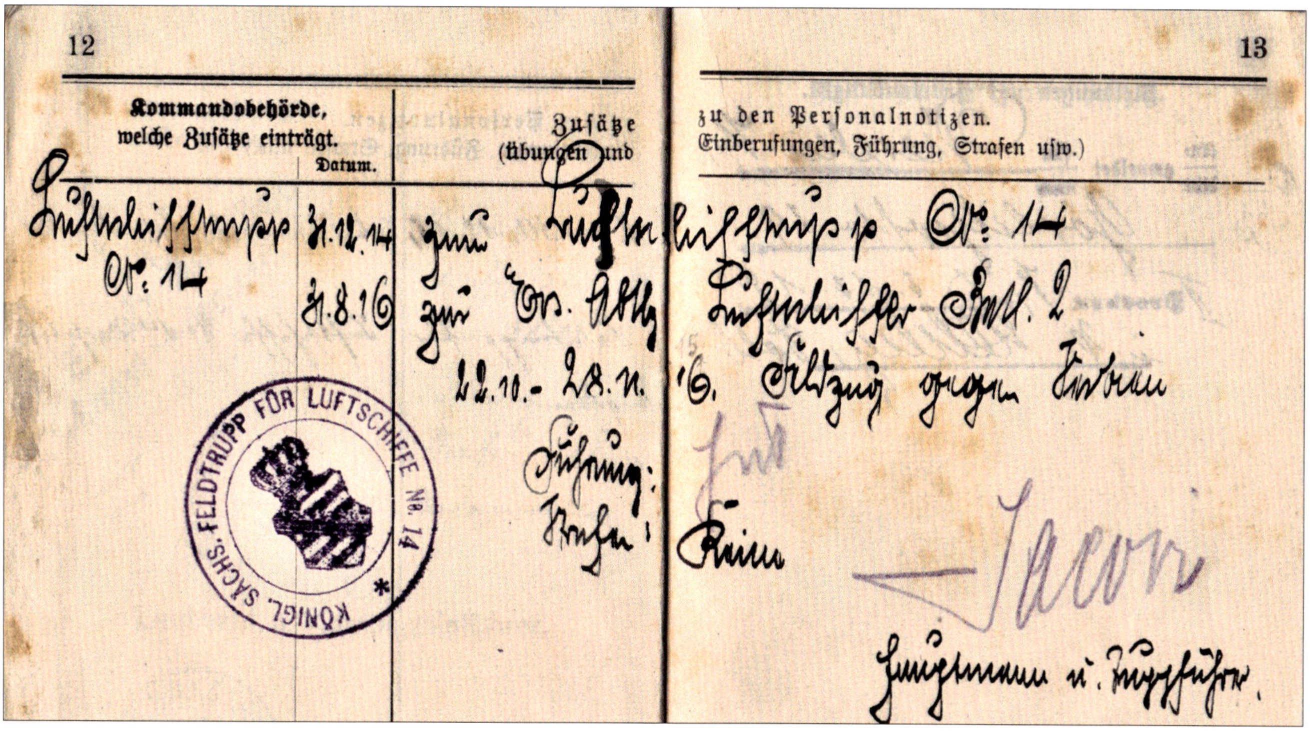
12
13
Kommandobehörde, welche Zusätze einträgt.
Datum.
Zusätze zu den Personalnotizen. (Übungen und Einberufungen, Führung, Strafen usw.)
Luftschiffertrupp Nr. 14
31.12.14 zum Luftschiffertrupp Nr. 14
31.8.16 zur Ers. Abtlg Luftschiffer Batl. 2
22.10.–28.11.16. Feldzug gegen Rumänien
Führung: gut
Strafen: keine
Jacobi
Hauptmann u. Truppführer.
KÖNIGL. SÄCHS. FELDTRUPP FÜR LUFTSCHIFFE NR. 14

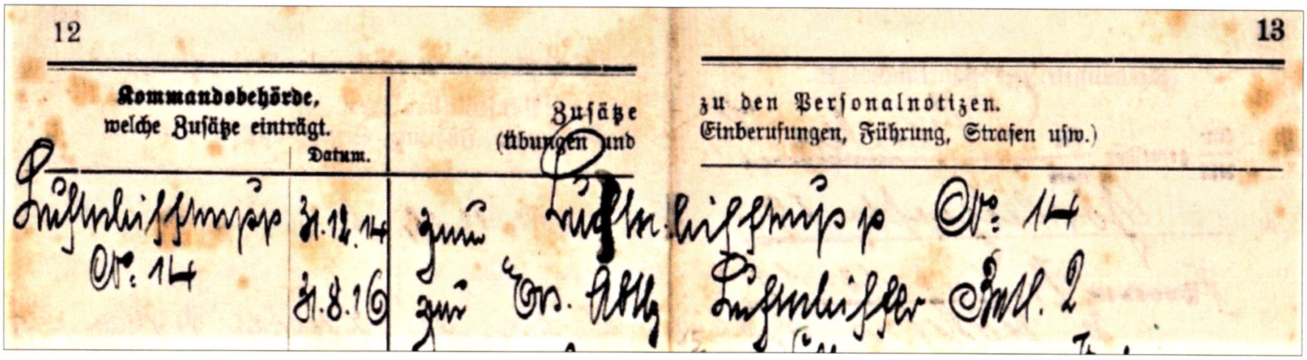
12
13
Kommandobehörde, welche Zusätze einträgt.
Datum.
Zusätze zu den Personalnotizen. (Übungen und Einberufungen, Führung, Strafen usw.)
Luftschiffertrupp Nr. 14
31.12.14 zum Luftschiffertrupp Nr. 14
31.8.16 zur Ers. Abtlg Luftschiffer Batl. 2

Luftschiffer-Ersatz-Abtlg. Nr. 2
29.8. 1916
zur nebenstehenden Abteilung und am 1.1.1917 zur 1. Seeflieger-Abteilung Holtenau versetzt.
Führung: gut
Strafen: keine
a. B.
Leutnant.
Luftschiffer-Ersatzabteilung Nr. 2 mit Rekrutendepot * Berlin *
WARNEMÜNDE

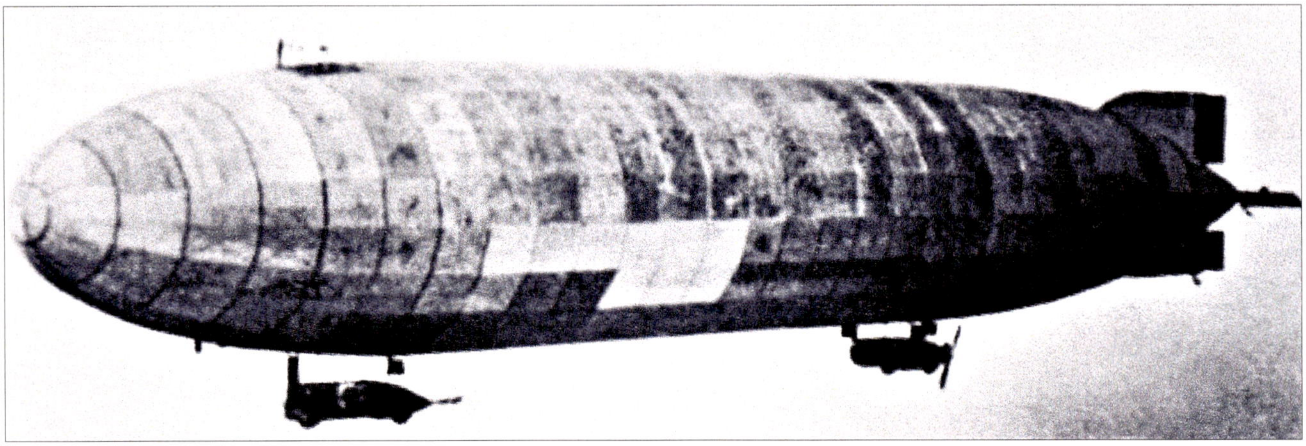

Image 200-3a Above: LZ 86, airborne (image: see scharekf.hu)

him and the entire "LT 14" Luftschifftrupp. Until they were officially notified of a change, however, they each went about their official duties.

6.2 Early August 1916 – "LZ 101" Stopover at Timisoara Airship Base

In late July it was decided that airship LZ 101 (LZ 71) would be stationed at Jamboli airship base in Bulgaria. Construction of the airship was completed on June 29, 1916. This airship was a q model; its main difference from the p airship models was that, from the beginning, the q model was equipped with a 35.000 m³ shell (see Griehl, 2010, page 80).

"On its journey to Jamboli airship base, LZ 101 made a stopover in Timisoara from August 1–3" (see Navy, 1990/I, page 8/**see image 197).** The airship landed at 3:45 on August 1. "At a length of 196.50 m, diameter of 23.90 m, volume of 56,000 m³, 1300 HP, and a speed of 114.48 km/h it considerably exceeded the performance of the previous airship constructions (LZ 01, etc.) (SANKTANDRES/1918, page 168).

While the stopover was not planned, it became necessary as LZ 101 had passed multiple storm fronts on its long journey to Jamboli which caused damage to the airship. Personnel of Königlich Sächsischer Feldtrupp für Luftschiffe Nr. 14 in Timisoara were able to repair the damage, allowing the airship to continue its trip to Jamboli airship base under the command of Hauptmann Gaissert on August 3, 1916.

6.3 "LT 14" Transfer to Wittmund (Lower Saxony)

"On August 12, 1916, deutscher Luftschiffer-Trupp 14 transferred from Temesvár to Wittmund (district of Aurich). This freed up several accommodations in Szentandrás which, with permission of the German Feldflugchef, were utilized to provide housing to the training troop (Schulkompanie). Since July 24, 1916, Ballonersatzkompagnie Nr. 2 was operating out of Szentandrás (at the time, Ballonersatzkompagnie was officially known as Luftschiffer-Ersatzkompanie). This was necessary because, as a result of strategic reconfigurations, Ballonersatzkompagnie Hainburg was unable to fund missions and staff listings on its own.

The goal of adding Ballon-Ersatzkompanie Nr. 2 to the existing Luftschiffer-Ersatzkompanie in Szentandrás was to provide improvement and relief to training operations" (Kustan & Penz, page 18).

The new deployment site for "LT 14" at Wittmund was located in the vicinity of Bremerhaven. Wittmund airship base was taken over by Marineluftschifftrupp Nr. 4 in 1917. Königlich Sächsischer Feldtrupp für Luftschiffe Nr. 14 (Imperial Saxon Field Forces Nr. 14) subsequently took over the support of Weiterstadt airship base near Darmstadt. "LT 14" deployed there in March of 1918 (see Stadtwiki Dresden).

Wittmund is located near the city of Bremerhafen. Paul Biber's presentiment of a significant change became a reality. Paul Biber was specifically requested to join 1. Seeflieger-Abteilung Holtenau to take on particular tasks based on his diverse professional knowledge and skills which he demonstrated while carrying out technical tasks with Königlich Sächsischer Feldtrupp für Luftschiffe Nr. 14, as well as his occupation as a technician with well-known Dresden-based measurement engineering company G. Rosenmüller.

Paul Biber left Königlich-Sächsischer Feldtrupp für Luftschiffe 14 ("LT 14") in late August (see images 198 and 199) to travel to the North Sea.

7. "LZ 86", "LZ 81", and "LZ 97" War Patrol Flights Departing Timisoara Airship Base Through Late December 1916

Image 200-1 Left: Successful return of LZ 101 after the bombing of Bucharest (image: Schmeelke Archive).
Image 200-2 Above: Albatros D.III supported the bombing of Bucharest (image: Schmeelke Archive)

7.1 Loss of "LZ 86" and "LZ 81" on Their First Offensive Flights

Following the relocation of Deutscher Luftschifftrupp 14, the German Feldflugchef put in a request with LFT command to continue operations to accommodate airships at "ADEBAR" by keeping Austrian-Hungarian personnel stationed there.

On August 27, 1916, Romania declared war on the Central Powers. The following day, Germany's declaration of war on Romania followed.

In the night of August 28, 1916, LZ 101 (see image 200-1) departed Jamboli to launch an attack on Bucharest (see Navy, 1990/I, page 8). From an altitude of 3500 m, 1.5 tons of bombs were dropped on military camps and the north railway station.

Later, a large bomb hit the gendarmerie barracks, crossing Romania's deployment plans. In the following days, German war aircraft finished the task that LZ 101 had swiftly started (see Lehmann, 2019).

Immediately following the reconstruction efforts in Dresden, (see image 200-3b) German airship "LZ 86" arrived at Timisoara airship base on August 30, 1916 (departing from Vienna) (see image 200-3a). "Members of Hauptmann Walter Wolff's crew (see image 200-3c) included Offiziere Brenner, Kramer, Köstlich...as well as 'Bobby', the dog." (SANKTANDRES/1981, page 169). (see image 200-3d)

On September 3, 1916, LZ 86 (see image 200-3e) departed on its first war patrol flight at the new theater of war to launch an attack on Bucharest and Ploesti. (see image 200-3f)

"Following a successful mission with a bomb drop of approximately 2000 kg overhead Ploesti, the airship reversed course and reached its home base the next day around noon. On approach toward the hangar at Szentandrás, the airship suddenly sagged; the forward nacelle forcefully impacted the ground near the village of Freidorf, causing it to tear off.

The occupants died on impact, among them Hauptmann Wolff, airship Kommandant, his three Offiziere, three Steuermänner (navigators), the Maschinist (machine operator), as well as the Funker-Unteroffizier (radio operator).

The airship, now unpowered, rose back into the air with the rear nacelle still attached. Approximately two hours later, it touched back down in a field between the towns of Utvin and Sankt Michel, 6 km west of Freidorf. The three surviving crew members who were still in the nacelle were largely unscathed" (Ploes, 2018, page 21) (see image 201a through 201F).

The unusual flight characteristics of "LZ 86" could be the consequence of a possible hit by new French air defense rockets overhead Ploesti (see

Image 200-3b Above: Modification of airship LZ 86 inside the airship hangar at Dresden-Kaditz, August 1916. (image: Fabry Archive)

Image 200-3e Above: Airship LZ 86 is moved out of the hangar at Timisoara on September 3, 1916, to be readied for the attack on Ploesti. (image: Fabry Archive)

Image 200-3d Above: Airship LZ 86 Commander Walter Wolff, other officers, as well as Bobby, the dog. (image: Fabry Archive)

Image 200-3c Left: Hauptmann Walter Wolff, Commander of airship LZ 86. (image: Fabry Archive)

Image 200-3f Below: Airship LZ 86 enroute to Ploesti, September 3, 1916. (image: Fabry Archive)

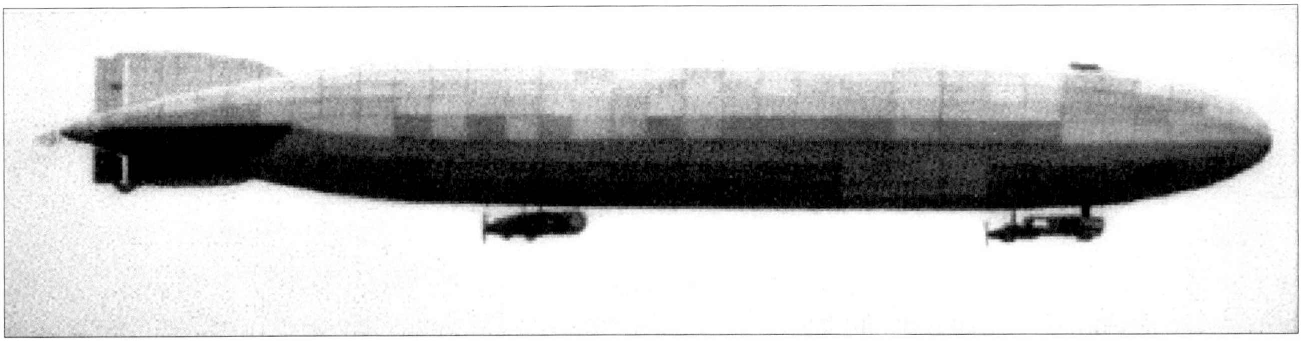

Image 201a Above: The front nacelle of airship "LZ 86" was torn off and destroyed upon impact with the ground (image: see scharekf.hu)

Image 201b Above: Airship LZ 86 wreckage (rear part), following its crash landing near Freidorf on September 4, 1916. (image: Fabry Archive)

Image 201c Above: Airship LZ 86 wreckage (middle part), following its crash landing near Freidorf on September 4, 1916. (image: Fabry Archive)

Image 201d Above: Airship LZ 86 wreckage (front part), following its crash landing near Freidorf on September 4, 1916. (image: Fabry Archive)

Image 201e Above: The torn off and destroyed nacelle of airship LZ 86 (including propeller), September 4, 1916. (image: Fabry Archive)

Image 201f Above: Torn off and destroyed front nacelle of airship LZ 86; its cab was completely shattered. September 4, 1986. (image: Fabry Archive)

Heilbronn, den 18. September 1916.

Danksagung.

Für die liebevollen Beweise herzlicher Teilnahme beim Tode meines einzigen Sohnes

Hauptmann Walther Wolff

sowie für die ehrenvolle Begleitung zu seiner letzten Ruhestätte spreche ich im Namen meiner Familie innigsten Dank aus.

Der Vater:

August Wolff, Oberleutnant d. L.

z. Zt. beim Kaiserlich deutschen Gouvernement Warschau.

Friedrichshafen, September 1916.

Image 201g Above: Expression of thanks for the participation in the airship LZ 86 memorial service for Walter Wolff, composed by his father, August Wolff (Neckar newspaper, September 19, 1916). (image: Fabry Archive)

Heilbronner Chronik.

Heilbronn, 7. September.

Heilbronner Ehrentafel.

Gefallen:

Walter Wolff, Hauptmann und Kommandant eines Luftschiffes; Inhaber des Eisernen Kreuzes 1. Klasse und anderer Auszeichnungen. — Walter Wolff ist am 9. Januar 1896 in Heilbronn geboren als Sohn des Fabrikdirektors Wolff, später in Friedrichshafen, z. Z. Oberleutnant d. Lw. beim Kais. Gouvernement Warschau. Er wurde am 27. Januar 1907 Leutnant und trat von hier aus am 1. Oktober 1913 in die 4. (württ.) Luftschifferkompagnie über, mit der er in die Garnison Friedrichshafen einzog. Seit Kriegsbeginn befand sich Hauptmann Wolff im Felde und hatte an vielen glücklich verlaufenen Fahrten unserer Luftschiffe reichen Anteil. Bei einer solchen hat er nun den Tod fürs Vaterland erlitten.

Image 201h Above: Obituary for LZ 86 the commander of LZ 86 in the Neckar newspaper, September 7, 1916, page 2. (image: Fabry Archive)

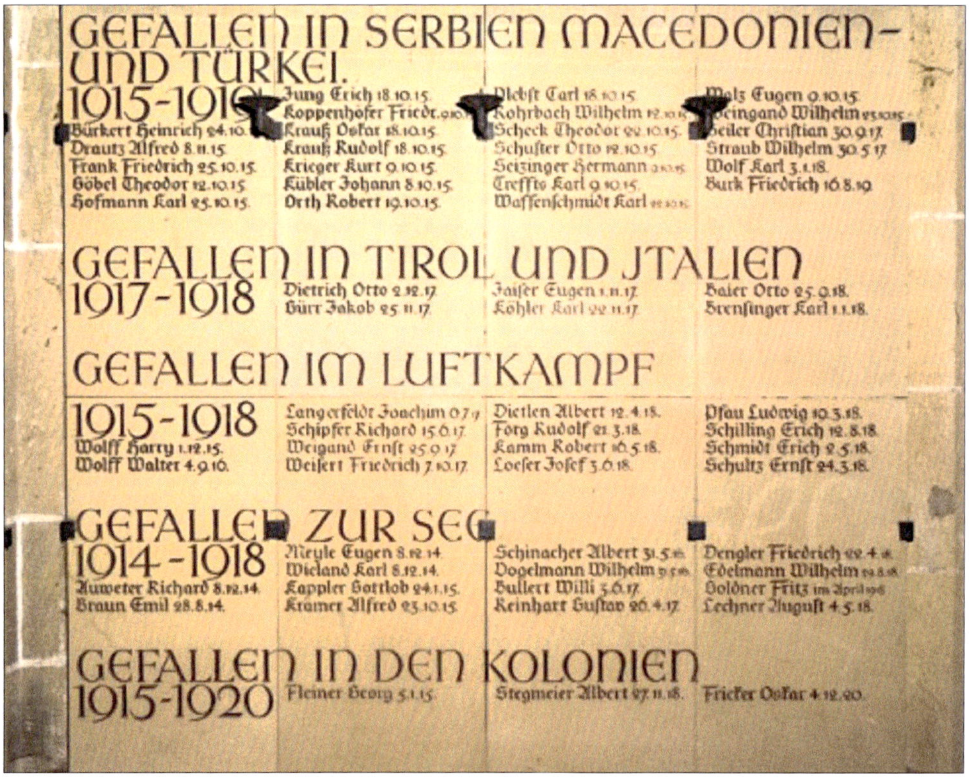

Image 201i Left: Walter Wolff is listed among the military members killed in aerial combat on the commemorative plaque dedicated to those who fell during WWI. The plaque is in the southern hall of Hafenmarktturm in the town of Heilbronn. (image: Fabry Archive)

Image 201j Above: The caskets with the remains of the aeronauts who perished in the crash of airship LZ 86 were laid out in Timisoara. September 7, 1916. (image: Fabry Archive)

Image 201k Above: The funeral procession proceeded toward the military cemetery where the perished aeronauts were laid to rest. September 7, 1916. (image: Fabry Archive)

Image 201l Above: The funeral procession with the caskets of the perished LZ 86 aeronauts slowly moved across the city of Timisoara. September 7, 1916. (image: Fabry Archive)

Image 201m Above: Funeral procession with the caskets of the perished LZ 86 aeronauts slowly moved toward the war cemetery where a formal burial was held with a large attendance, September 7, 1916. (image: Fabry Archive)

Image 201n Above: Original gravestones and memorial at the Timisoara war cemetery. (image: Fabry Archive)

Image 202 Left: Gravestone commemorating the Luftschiffer killed in the accident of "LZ 86", Timisoara cemetery (image: Nelepcu, 2020).

ADZ, 5-2014).

The mortal remains of Kapitän Walter Wolff and his officers, Brenner, Kramer, and Köstlich, were transferred to their homeland where they were laid to rest. The rest of the fallen crew members were laid to rest at the Timisoara cemetery on Lippaer Straße following a large funeral procession **(see images 202, 203, and 204).**

An inscription on the gravestone at Timisoara cemetery lists the names of fallen crew members Funkunteroffizier Otto Schienkler, Unt.-Steuermann Karl Reimer, Unter-Maschinist Jacob Renz, Ob. Steuermann Franz Thomsen, and Unt. Steuermann Georg Trüb. Beside them is the final resting place of Ob. Steuermann Max Schmidt who died during a Zeppelin mission on March 2, 1916" (see ADZ,

Image 203 Above: Inscription in the base of the gravestone for "Luftschiff LZ 86" crew members killed in the accident, Timisoara cemetery (image: Nelepcu, 2020).

5-2014).

The loss of "LZ 86" considerably restricted the scope of action for German airships against Romania. As a result, airship "LZ 101" which was stationed at Jamboli airship base in Bulgaria was made available for war patrol flights against Romania.

After "LZ 86" was no longer available following its crash landing, on September 4, 1916, efforts were immediately directed to finding a suitable replacement. German military command recognized the seriousness of the situation and requested the deployment of successful Army airship "LZ 81" to Timisoara.

In November 1915, "LZ 81" had established airship operations in the Balkans from Timisoara airship base. In early 1916, "LZ 81" left Timisioara and successfully participated in multiple operations on the western front originating from stations in Namur and Düsseldorf (see image 205). From there it was subsequently removed and deployed to Dresden.

While at Dresden-Kaditz airship base, the airship's conversion to a "q" variant took place in June and July, extending its length by 15 meters (from 163.5 m to 178.5m) which also increased its volume from 32,000 m^3 to 35,800 m^3 (see Griehl, 2010, pages 78–80).

Timisoara airship base commander, Hauptmann Kornél Poppe was an enthusiastic balloon and Zeppelin operator. While in Timisoara, Hauptmann

UNT. MASCHINIST
JAKOB RENZ
4.IX.1916
OB. STEUERMANN
FRANZ THOMSEN
UNT. STEUERMANN
GEORG TRÜB

OB. STEUERMANN
MAX. SCHMIDT
2.III.1916

UNT. STEUERMANN
GEORG TRÜB
4.IX.1916

UNT. STEUERMANN
KARL REIMER
4.IX.1916

OB. STEUERMANN
FRANZ THOMSEN
4.IX.1916

Image 204 Facing Page, Above, & Right: Gravestones for each of the "LZ 86" Luftschiffer who perished in the crash, Timisoara cemetery (image: Nelepcu, 2020).

Image 205 Below: Photo card (LZ 81 on the western front) from LZ-81 Steuermann (navigator) Reuter (Reuter, 1916)

Image 206 Facing Page: Balloon rising near Timisoara airship hangar (image: see scharekf.hu)

Image 207-1 Above: Austro-Hungarian Flieger arrive for refueling at Timisoara airship base (image: see scharekf.hu)

Image 207-2 Above: Short fuel stop with immediate continuation of the flight. The "wooden airstairs" to access the aircraft are particularly noteworthy. (image: Racil Archive)

Poppe supported the training of balloonists as well as the utilization of balloons for the collection of weather data at various altitudes along with determining the height of cloud or fog layers overhead the Timisoara target region. Occasionally, Hauptmann Poppe would get into the balloon basket himself (see image 206).

While Timisoara airship base was not operating an airship at the time, it was frequently utilized as a point of landing for aircraft on stopovers on their way to destinations further east, such as Jamboli, for example (see image 207-1 through 207-4).

There is a message from the commander of Luftschiff-Trupp 14 Dresden-Kaditz, dated September 12, 1916, regarding LZ 81's arrival in Dresden (archival document Nr. 1033)

Image 207-3 Above: Following a successful landing in the fall of 1916, the pilot and his mechanic spent the night at Timisoara airship base. "Schiffspflegetrupp" personnel take care of aircraft maintenance and refueling. (image: Racil Archive).

Image 207-4 Above: The aircraft is ready for startup. The propeller was cranked by hand. Once the engine reached the proper RPM, its tail lifted up and ground personnel turned the aircraft in a suitable position for take-off. The rocks (chocks) were then removed using a rope, allowing the aircraft to start moving across the surface (image: Racil Archive).

In mid-September, "LZ 81" departed Dresden under the command of Hauptmann Barth. Its route of flight took the airship to the Balkans operating area via Vienna. It safely landed at Timisoara airship base, providing the base with an operational airship once again (see image 208).

Right away, preparations began for the airship's first war patrol flight (see image 209). In addition to the long distance, the desired course to Bucharest and the routing across mountainous terrain posed challenges for the trip, particularly due to the prevailing wind direction overhead the Danube. The river Danube forms the border between Bulgaria and Romania. The airship had to take care to avoid shots fired by Romanian cannons stationed in the border area.

In an attempt to avoid enemy fire, airships climbed to higher altitudes which led to some crew members suffering from shortness of breath. As a result, airships were partly equipped with breathing equipment. Breathing with the assistance of such apparatuses (see image 210) was essential for survival; it was practiced routinely.

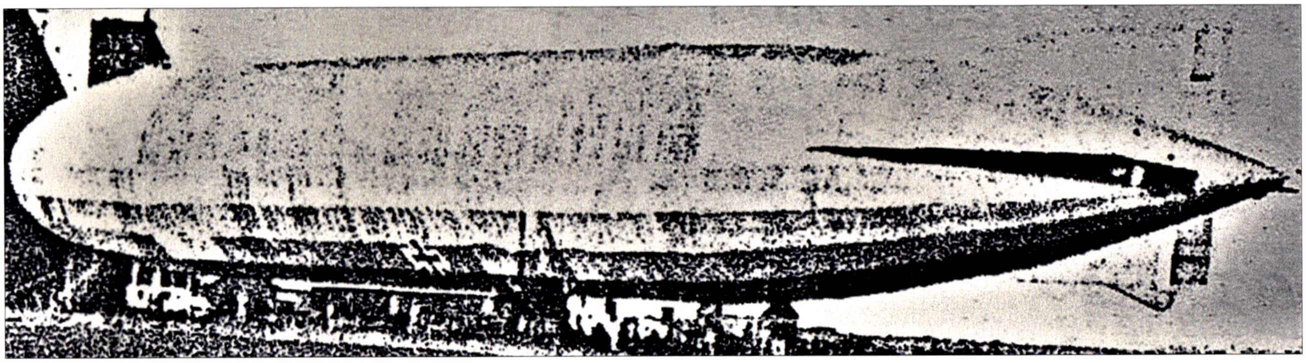

Image 208A&B Above & Top: The extended version of LZ 81 returned to Timisoara (image: Navy, 1990/I, page 9).

Image 209 Below: LZ 81 Travel route from Timisoara to Bucharest (using a Navy chart, 1990/I, page 7).

Image 210 Facing Page: Crew members practicing the use of breathing apparatuses (image: Racil Collection)

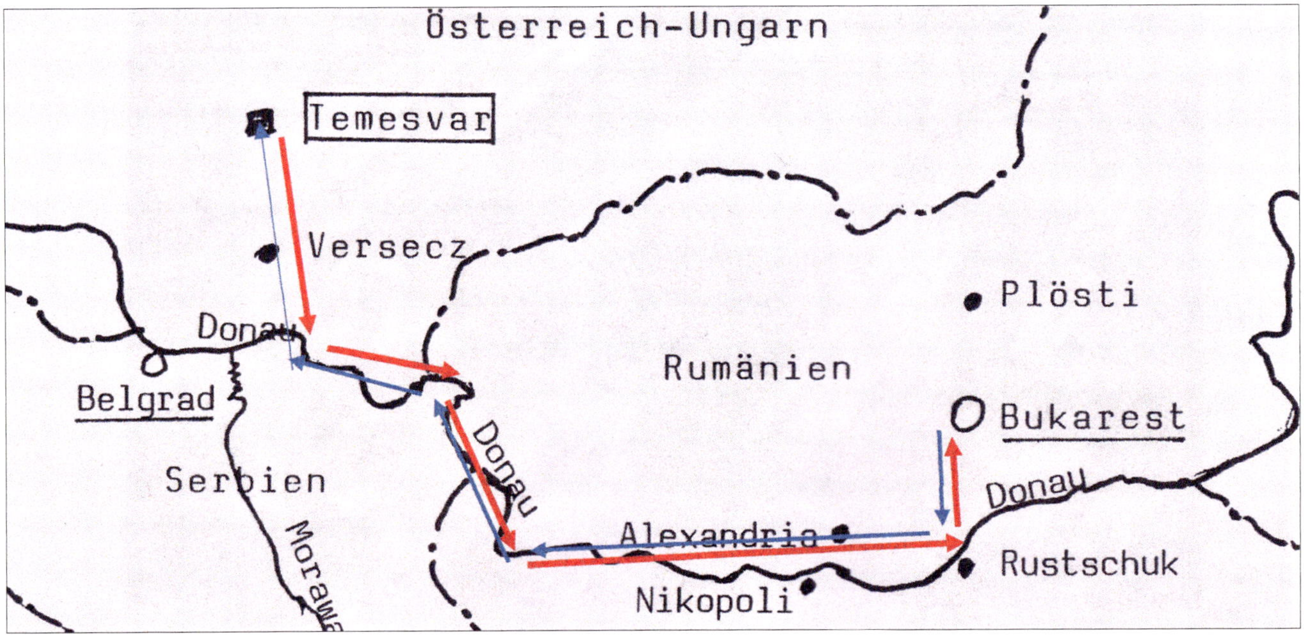

Image 211 Above: Landungstrupp gathering in front of the Timisoara airship hangar (image: see scharekf.hu)

Image 212 Facing Page: "LZ 81" overhead Bucharest, September of 1916 (see Post LZ 81)

On September 18, 1916, k.u.k. War Office ordered LFT command to withdraw the Ballonmannschaft from Timisoara Luftschiffterersatzkompagnie in order to merge it with the local Ballonmannschaft at Hainburg. Ballonersatzkompagnie was responsible for training of Ballonmannschaft while Luftschiffterersatzkompagnie was solely responsible for Luftschiffer services (see Kustan & Penz, pages 18/19).

On September 24, "Landungstrupp" gathered in front of the airship hangar, awaiting orders for "Aushallen" of "LZ 81" (move airship out of the hangar) (see image 211).

Following "Aushallen", orders were swiftly issued to raise the airship ("Hoch!"). LZ 81 rose up, turning on a southerly course towards the Danube. Once the river was reached, the airship's route followed the Danube.

"Shortly before reaching Bucharest, LZ 81 turned away from the river and altered its course towards the city, reaching its destination in the middle of the night. Bucharest's city lights were visible from a great distance; as soon as the airship was spotted however, all city lights were turned off. Only the fire of the air defense shots was visible as shooters marked their target. The attack was a success nonetheless, all bombs were dropped on the Bucharest stronghold...LZ 81 reached its home base on the same day, nearly undamaged. Preparations began right away to ready the airship for its next mission." (Ploes, 2018, page 22)

This event was impressive in nature though frightening for the population of Bucharest. The threat to Bucharest and the bombing of the city by airships was even captured in a postcard (see image 212).

7.2 September 1916 – Loss of "LZ 81"

Immediately following its landing, functionality checks were carried out on "LZ 81"; bullet holes were repaired, hydrogen was replenished, and ammunition brought onboard. Troops were working tirelessly to ready the airship for its next mission. The following day, on September 26, "LZ 81" departed under the command of Hauptmann Hans Barth for a second bombardment of Bucharest (see

Zepelinul deasupra Bucureştului.

Image 213-1 Above: "LZ 81" in flight towards Bucharest (image: Biber Archive).

Image 213-2 Above: English war ships using spotlights in search of an airship (image: Schmeelke Archive).

Image 213-3 Right: Spotlights were transported on cars to help search for the airship (image: Schmeelke Archive).

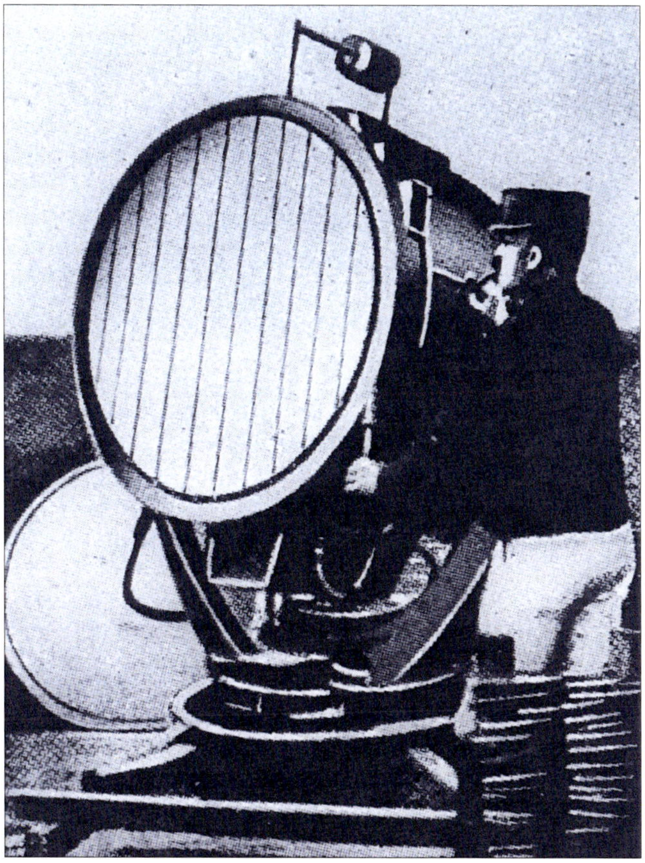

Image 213-4 Left: Some spotlights were large versions from England (image: Schmeelke Archive).

Image 214 Below: LZ 81 flight barogram from September 26/27, 1916 (see scharekf.hu)

Image 217-1 Above: L.Z. 97 command seal stamp (see scharekf.hu)

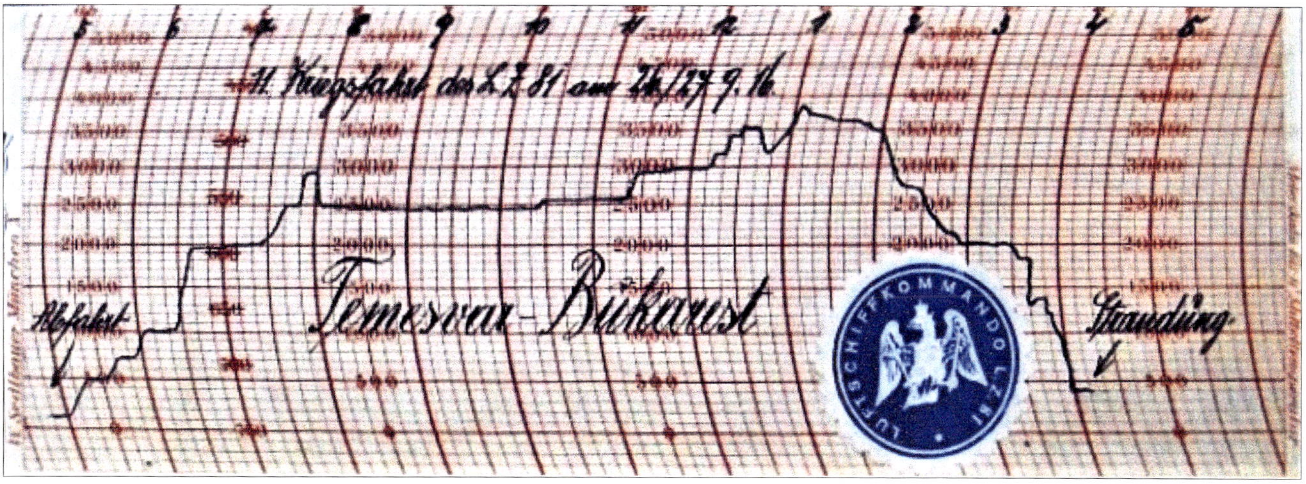

image 213-1).

"Weather conditions steadily improved compared to the trip carried out on September 24. Wind strength decreased, resulting in the departure time being delayed until 17:00. In order to pass overhead Bucharest in a north-south direction in prevailing northerly winds, the airship crossed the Danube near Nikopoli. As soon as the airship had crossed the river, it began experiencing heavy fire. By the time LZ 81 had reached Alexandria, the lights of Bucharest's northern port were clearly visible. As soon as the airship passed overhead Alexandria however, the city lights were extinguished to conceal the view of Bucharest once again. The sudden extinguishing of lights must have been the result of systematically arranged reporting procedures.
After passing Alexandria, the airship turned on course to head north of Bucharest. When the airship crew determined their position north of the city (using dead-reckoning procedures), they turned southbound. In that moment, the port engine quit as a result of a connecting rod failure. The crew was

Image 215 Above: LZ 81 crash landed near Tirnova (Bulgaria) (image: see scharekf.hu)

unable to restart the engine for the remainder of the flight. Shortly after, the city's first spotlights lit up; spotlights of the entire city soon followed suit.

Floodlights were steadily pointed at the airship (see images 213-2 through 213-4) while numerous air defense batteries opened heavy fire. Nevertheless, the airship passed overhead the city center where it dropped its entire load of 1650 kg explosive bombs and incendiaries.

LZ 81 experienced heavy fire long after passing the city center, Bucharest's spotlights still locked on the airship. Crew members reported two larger fires in Bucharest as viewed from the airship even after passing the Danube. According to the accounts of pilots who overflew Bucharest the morning of September 27, the fires were still burning at that time" (see Navy, 1990/I, pages 9/10).

The barogram shows some interesting data regarding the flight of LZ 81 **(see image 214).**

"Even before the airship reached the river Danube, crew members noticed its increasingly inclined position. After the river crossing near Rustschuk it was discovered that the number 15 and 16 cells were empty; two other cells showed signs of hydrogen loss. The airship must have received two significant hits along with several smaller ones. Four gas tanks were thrown overboard along with spare parts and two MGs in order to lighten the airship's load. As a result of the fuel tanks being dumped, the airship was no longer able to reach Timisoara on the remaining supplies. The decision was made to fly to Jamboli via Tirnova. An emergency signal was broadcast to the respective FT station at 02:45.

As a landing at Jamboli seemed impossible if the airship continued to fall flat, Tirnova was advised of the intention to land there should it become necessary. 6 km prior to reaching Tirnova, the airship suddenly started descending at a high rate, making it impossible to control. The crew succeeded in maneuvering the airship into the wind and landing it in a brush covered area **(see image 215).**

The forward nacelle impacted first; it was separated from the hull and completely destroyed. The airship hull was also destroyed. The aft part

Image 216 Above: "Crew of airship "LZ 97" (image: see scharekf.hu)

section of the airship rose back into the air about 50 meters before touching down intact after the crew opened all the valves" (see Navy, 1990/I, pages 10/11).

Miraculously, only two crew members were seriously injured in the hard landing of the airship.

"The injured crew members were immediately taken to the 211th field hospital at Gorna-Orchowitza where Unteroffizier Esenwein succumbed to his injuries at 16:15. Hauptmann

Barth was the other person to sustain injuries. He suffered a dislocated left shoulder and stayed at the 211th field hospital until his transfer to field hospital 122 in Timisoara on October 7" (Ploes, 2018, page 22).

While "LZ 81" successfully fulfilled its mission to "bomb Bucharest", the airship sustained several hits as a result of the attack which significantly impacted its flying and maneuvering abilities and ultimately led to the crash landing in Tirnova. The previously very successful airship was badly damaged; it was ultimately dismantled and wrecked due to a lack of spare parts and as a result of high expenditure.

"Hauptmann Kornel Poppe also participated in the war patrol flight on September 26/27, 1916. On November 4, 1916, he notified the Luftschifferkommando in Vienna that he had been presented with Eisernes Kreuz II. Klasse (Iron Cross 2nd Class) by GFM von Mackensen for that flight." (Kustan & Penz, page 19) (GFM= Generalfeldmarschall, field marshal).

Bucharest, and particularly its stronghold, were bombed intently between September 25 and 27, not only by Zeppelin airships "LZ 81" (2 flights originating from Timisoara, carrying a total of 4,513 kg of bombs), "LZ 101" (originating from Jamboli/Bulgaria), as well as German fighter aircraft from O.H.L. (Kagohl) Kampfgeschwader 1 (fighter squadron 1) (see Ploes, 2018, page 22).

7.3 October – "LZ 97" Stationing at Timisoara Airship Base

Once again, a void caused by the second destroyed airship had to be filled in the Balkans theater of war. In search of a suitable replacement, "LZ 97" was chosen as the replacement airship. "LZ 97" took its maiden voyage on April 4, 1916. On two occasions, the airship had dropped its bombs on London and Boulogne (see ADZ, 5-2014).

Following the latest successful attack on Boulogne on September 22, 1916, originating from its home base at Namur, "LZ 97" Kommandant Oberleutnant Helmut Weidling and his crew **(see images 216 and 217-1)** received orders to deploy to Timisoara airship base (see Ploes, 2018, page 22).

"On September 25, 1916, AOK entrusted Kommando der Luftfahrtruppen with the establishment of a stable field weather station at Timisoara. In military and disciplinary terms, employees of the field weather station were subordinate to Kommando der Luftschiffer-Ersatzkompagnie Szentandrás while concerns

Image 217-2 Above: Bartsch-Sigsfeld type Drachenballon (image: Mückler, 2013, page 131).

affecting field weather services were reported directly to LFT command. On October 19, 1916, Dr. Alfred Roschkott who had been associated with the German field weather station in Timisoara and returned to Vienna, traveled back to Timisoara to take command of the Timisoara fixed field weather station.[17]" (Kustan & Penz, pages 19/20).

"K.u.k. troops were deployed to another German unit stationed in Szentandrás. In order to gain knowledge in the operation of weather "kites" (balloons), Korp. Josef Haar and Zgsf. Karl Hutter transferred to the German Felddrachenstation in Szentandrás on October 9, 1916. They had been assigned to Luftschiffer-Ersatzkompagnie. Following their return to Vienna, Drachenversuchsabteilung Fischamend was decommissioned. On October 21, 1916, the first k.u.k. Drachenstation[18] was established in Straßhof near the northern railway. It was relocated to Ritten near Bozen in early 1917." (Kustan & Penz, page 20)

"Drachenballone" were elongated balloons shaped like cylinders with funnel-shaped vestibules developed in 1883 by engineer August von Parseval and Prussian Hauptmann d.R. Hans Bartsch von Sigsfeld (see Mückler, page 124). (see images 217-2 & 217-3) "Drachenballone" had increasingly streamlined shells and a directionally stabilizing "tail" which was superior to spherical balloons in windy conditions. "Drachenballone" were utilized as a heightened observation platform to monitor combat and artillery. The Navy also utilized

[17] Remark: the following field weather stations were located on the northeastern front: FWSt. Lublin station; FWSt. Kielce station; FWSt.1 station near Flik 10 in Wladimir Wolinsky; FWSt.2 station near Flik 14 in Krasne; FWZ station in Lemberg; FWSt.11 near Flik 11 in Slope; FWSt.13 near Flik 7 in Chocin; FWST.3 near Flik 1 in Besztercze; FWSt.4 near Flik 31 in Gyula-Fahervár; Stab FWSt. Temesvár, new; FWZ Budapest, new.

[18] Remark: The Drachenstation was part of field weather service. Its main purpose was to use kites to determine air properties and identify currents in different layers of air at various altitudes. This became necessary as aircraft continually reached greater altitudes.

Image 217-3 Above: "Reproduction of an English AE-Drachenballon (from 1917); its shell was now streamlined, and its "tail" increased toward the stabilizing rear fin (image: Mückler, 2013, page 131).

"Drachenballone". They were also utilized by destroyers to locate and destruct submarines. The Navy and Army transmitted reconnaissance results via telephone to the command center.

On October 17, 1916, LZ 97 arrived at Timisoara airship base in southern Hungary under the command of Oberleutnant Helmut Weidling as a replacement for LZ 81. It departed on its first war patrol flight toward Bucharest on October 23/24.

Upon its safe return from a successful bombing of Bucharest, LZ 97 was swiftly prepared for its next war patrol flight.

The following day, airship LZ 97 bombed the railway junctions Ciulnita, Fetesti, and Ploesti, supported by airship LZ 101 which was stationed in Jamboli (Bulgaria), the second German airship base in the Balkans (see: Navy, 1990/I, page 11).

In late October 1916 (October 23/24), "LZ 97" (based at Szentandrás-Temeswar) and "LZ 101" (based at Jamboli) bombed Bucharest together. "On December 6, 1916, Bucharest was finally taken by German army units" (Kustan & Penz, page 19).

While stationed at Jamboli, "LZ 101" carried out other successful war patrol flights, for example to Galatz, Mudros, and Odessa.

An attack on Galatz (Romania) was launched the night of December 25, 1916:

"At 6:30 pm an ascent was carried out, heading toward Varna...at 10:50, the skies lit up more due to visual contact with a larger, brightly lit town near a lake while the airship flew a loop. The town and lake were identified as Babadag and Lac Racin, respectively.

Upon reaching the Danube, weather conditions were almost clear. 10km west of Tulcea, the Danube was crossed. 15km past the river, the airship turned on a westerly course where it encountered strong winds. At slow speeds, the airship passed by the well-lit town of Reni while maintaining its course toward the western shore of Lacul Brates. In spite of hazy conditions, the railway tracks leading towards Galatz were clearly visible. The airship followed the railway tracks to Galatz until the seaport lights were clearly identifiable.

Overhead the city, the airship altered its course from the south to the east; at 02:05 (Eastern European Time), the railway station and port facilities were overflown. The airship dropped 12x58kg explosive bombs on each target. Following the attack, the airship departed the area on a southeasterly course. 20 minutes later, it turned toward a compass heading of south-southeast." (see: Navy, 1989/IV, page 3)

8. The End of the War – A Tentative Ending of the Airship Era

Image 218 Above: LZ 101 airborne (image: see scharekf.hu)

8.1 Final Airship Missions in the Balkans

After the occupation of Romania in late December 1916, there were no further missions for "LZ 97" to attack Romanian targets. "LZ 97" was subsequently tasked with war patrol flights on Italian harbors (Valona, Brindisi, and Taranto). These long-distance flights were hampered by poor weather conditions; they were navigationally successful but not militarily (see Kustan & Penz, page 19).

"LZ 97 went on its last, verified war patrol flight on April 25, 1917, to Valona. The flight had to be discontinued like the previous missions, due to bad weather." (Ploes, 2018, pages 22/23)

Following the end of the mission, Army airship "LZ 97" departed Timisoara for Vienna on May 5, 1917. According to headlines in "Nachrichten vom Viertel unter dem Mannhartsberg", the airship appeared above the Korneuburg parade grounds at 08:00, accompanied by an aircraft, while on its return to Jüterbog (Germany). On July 5, 1917, it was decommissioned and disarmed (see Kustan & Penz, pages 20/21 and Ploes, 2018, pages 22/23).

Upon the termination of Army airship aviation at the end of April, Timisoara airship base was abandoned. A new occupancy of the airship hangar by another airship was not planned.

Jamboli airship base on the other hand was still in operation, as airship "LZ 101" (see image 218) was still stationed there at the time. The airship was tasked with attacking English footholds on the Aegean Islands and Odessa. "LZ 101's attempts to attack Odessa, Mytiöene, and Jassy failed. In the summer of 1917, an attack on Mudros was successful (...). Jamboli airship base was taken over by the Navy" (see Schwaebische Zeitung, November

Image 219 Above: Code name "ADEBAR" airship base (image: Paul Biber).

Image 220-1 Above: Departure of Africa-Airship L 59 (LZ 104) at Jamboli (Bulgaria), heading for German East Africa. (Griehl, 2010, page 92)

14, 2015)

After successfully dropping 11,934 kg of bombs during its war patrol flight missions, LZ 101 left Jamboli airship base. The airship was scrapped in September 1917.

"As the German field weather station in Timisoara had been moved to Sofia in the meantime, the necessity arose to establish a new k.u.k. field weather station for the eastern portions of Hungary. This weather station was established in Budapest. In terms of military and disciplinary connections, the station was subordinate to Budapest military command, while field weather service concerns were to be addressed directly with Navy and Air Force command." (Kustan & Penz, page 19).

"Luftschiffer-Ersatztruppe was divided into the staff department, Luftschiffer-Ersatzkompagnie, Luftschiffer-Schulkompagnie, and Luftschifferwerft

Image 220-2 Above: Zeppelin works at Staaken, as viewed from an airship (image: Schmeelke Archive).

Image 220-3 Below: The airship Haltetrupp awaits the landing of "L 59" and the dropping of the retaining ropes (image: Schmeelke Archive).

Image 220-4 Above: Einhallen (moving airship into the hangar) of "LZ 59" into the airship hangar at Jamboli (image: Schmeelke Archive).

Image 220-5 Below: Aushallen (moving airship out of the hangar) of Navy airship "L 59" in Jamboli (image: Schmeelke Archive).

Image 220-7 Above: L 59 starting its Africa mission. (Image: Schmeelke Archive)

Image 220-8 Left: Navy airship "L 59" ready for departure, awaiting the "Schiff hoch!" command (image: Schmeelke Archive).

Image 220-9 Facing Page Above Left: "L 59" Navy airship commander, Kapitänleutnant Bockholt was in charge of the mission trip to Africa (image: Schmeelke Archive).

Image 220-6 Below: Navy airship "L 59" was transpoted to the departure point on the field (image: Schmeelke Archive).

Image 220-10 Facing Page Below: Weather chart utilized for flight planning of the airship's mission to Africa (image: Schmeelke Archive).

Image 220-11 Facing Page Above Right: Memorial for Kapitän Bockholt and his crew commemorizing their mission to Africa on airship "L 59" (image: Schmeelke Archive).

Kapitänleutnant Bockholt, Führer von L 59

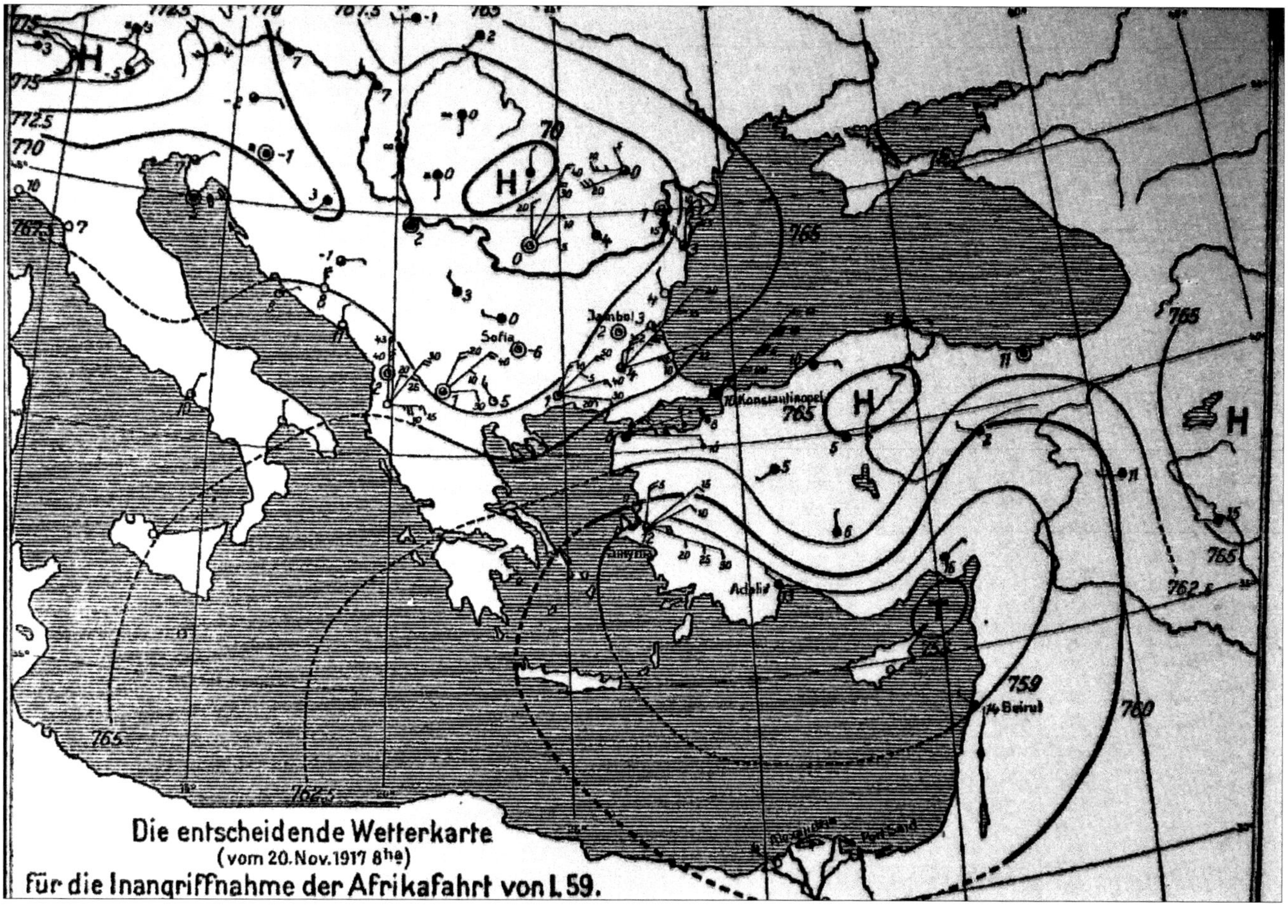

Die entscheidende Wetterkarte
(vom 20. Nov. 1917 8ʰᵉ)
für die Inangriffnahme der Afrikafahrt von L 59.

Image 221 Above: Closed hangar doors on the airship hangar in Temesvár (image: Biber Archive).

Image 222 Above: Out of use workshops at Timisoara airship base (image: Biber Archive).

(shipyard). Luftschiffer-Ersatzkompagnien (LSEK) were responsible for the practical training of Luftschiffermannschaft, replacement of Ballonabteilung departures, as well as the establishment of new formations. Luftschiffer-Ersatzkompagnien could be tasked with executing special Luftschiffer service tasks as needed.

On May 11, 1917, Luftschiffer-Ersatztruppe was established as an independent department while the staff department was assigned to LSET. The division of Luftfahrtruppe replacements was comprised of Luftschiffer-Ersatzkompagnie Nr. 1 in Hainburg on the Danube and Luftschiffer-Ersatzkompagnie Nr. in Timisoara at Szentandrás airship base." (Kustan & Penz, pages 20/21)

"On orders of kommandierender General der Luftstreitkräfte Thomsen, Luftschiffertrupp 23 which had previously been stationed at Timisoara was withdrawn in early May of 1917. The previous Truppenführer, Rittmeister d. L. Siemens, the Hafeninspektor, the Schirrmeister as well as 1 Unteroffizier and 6 soldiers remained at Timisoara to manage the base. According to a statement by the commanding General of the Air Force on May 20, 1917, there were no plans for an airship to be stationed at Timisoara airship base in the foreseeable future **(see image 219)**.

Szentandrás was utilized for hot air balloon rides by k.u.k. troops. Lt. Papsch, for example, received his hot air balloon pilot certificate on July 8, 1917." (Kustan & Penz, page 21)

Image 223 Above: Out of use accommodations at Timisoara airship base (image: Biber Archive).

Image 224-1 Above: Graves of French soldiers at the "Lippaer Straße" cemetary in Timisoara (image: Nelepcu, 2020).

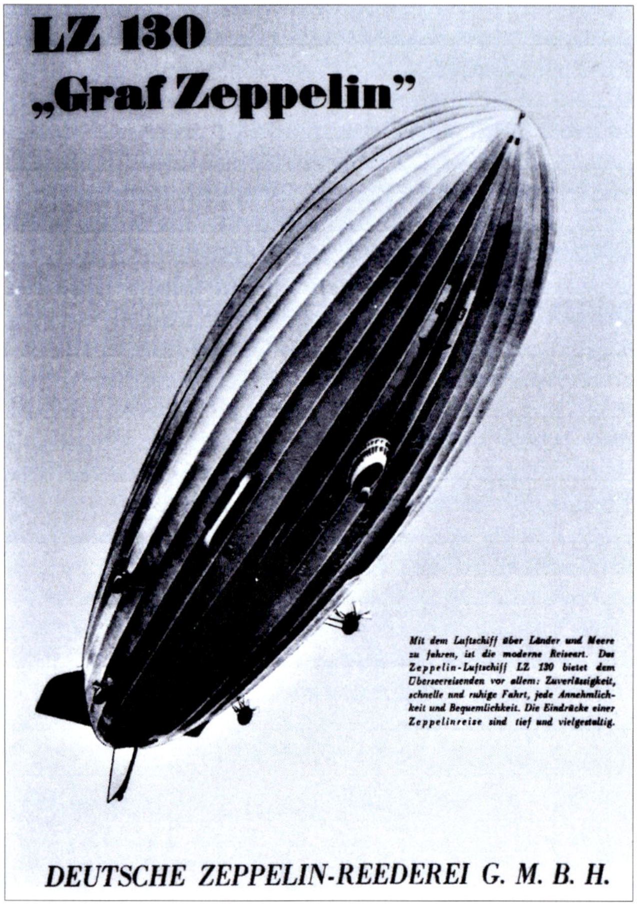

Image 224-2 Above: Travel to far away destinations was possible onboard airship LZ 130 *"Graf Zeppelin II"* (image: Schmeelke Archive).

"At a later time, disagreements arose between German Kommandant der Luftschiffhafenverwaltung Temesvár, Rittmeister d. Landwehr Siemens and Austrian-Hungarian Kommandant of k.u.k. Luftschifferersatzkompagnie Nr. 2, Hptm. Kornel Poppe. To clarify order sovereignty, a military command was established at Szentandrás on June 10, 1917, which also encompassed the airship base. The previously appointed base commander remained in command of the airship base. To prevent further "incidents" of this nature, Hptm. Kornel Poppe was removed from his command and transferred to a Ballonabteilung at VII. Korps (Isonzofront).

On July 5, 1917, Hptm. Ferdinand Baumann took command of Luftschifferersatzkompagnie Nr. 2 at Szentandrás.

As German Army Command had no further use for the airship hangar at Szentandrás (Temesvár), it was offered to the k.u.k. War Office for sale at a discount due to attrition. Until the end of the war the hangar was utilized by Luftschiffer-Ersatzkompagnie Nr. 2 under the command of

Image 224-3 Above: Airship *"Graf Zeppelin II"*, anchored inside the hangar (image: Schmeelke Archive).

Hptm. Baumann." (Kustan & Penz, page 22)

The new Navy airship L 59 (LZ 104) was stationed at Jamboli Naval air base. LZ 104 was a w type airship, equipped with the largest shell of any Zeppelin airships built in WWI. Its volume capacity was 68,500 m^3 (length 226,5 m / useful load 52,000 kg / max ceiling 8,200 m / range 16,000 km / number of crew 26) LZ 104 was built at Staaken; the airship had its maiden voyage on October 10, 1917. The Navy took over the airship as L 59 (Griehl, 2010, page 92).

On November 21, 1917, Navy airship L 59 departed on a long-distance flight to Africa on its third attempt (see images 220-5 through 220-10). The airship was to support General Letow-Vorbecks's troops in German East Africa (a German colony in the African Great Lakes region, which included what are now Burundi, Rwanda, and the mainland part of present Tanzania) with supplies. The distance from Jamboli to the Mahenge region measured 5,800 km.

"As there was no airship base in Eastern Africa, the airship was not able to depart again and return to Europe. The plan was therefore to dismantle the airship on-site to reuse the valuable component materials. The shell was to be transformed into waterproof sleeping bags and tarpaulins, the frame was to be used for the construction of radio masts or stretchers. The engines were used for the operation of radio stations." (ZEPPELIN MUSEUM BLOG 2017).

Overhead Smyrna (on the western coast of Turkey) and Kos, L 59 reached the African coast near Ras Bulau at 05:15 the following morning.

"L 59 referenced the river Nile for navigation as it continued heading southbound. (...) After leaving Egypt, the airship crossed into the Sudan. (...) In the early morning hours of November 23, the airship received a radio message from the Nauen radio station which issued orders to abort the mission and to return to Jamboli (see ZEPPELIN MUSEUM BLOG 2017).

"The crew of L 59 was deeply disappointed as they turned the airship around at 02:50 abeam Kharttum. (...) On the morning of November 25, the crew landed at Jamboli, completely overtired. (...) At the time of landing, L 59 had traveled 6,757 km in 95 hours, a new track record for a Zeppelin airship. L 59 had enough fuel in its tanks for another 64 hours of flight which would have allowed the airship to reach its destination" (ZEPPELIN MUSEUM BLOG 2017). A monument was erected in commemoration of airship "L 59"'s successful flight to Africa (see image 220-11).

Upon its return from Jamboli to Germany, L 59 was converted to a reconnaissance and attack airship.

"In the first months of 1918, the airship made a handful of valuable, far-reaching reconnaissance trips in the Mediterranean and two war patrol flights, one to Naples and one against Malta, the latter of which it did not return from. At the time when L. 59 was located approximately overhead the Strait of Otranto, a German submarine spotted a bright glow of fire in the sky. A fuel tank belonging to L. 59 was later found, leading to the conclusion that L. 59 must have crashed in the vicinity. The cause was assumed to be a gasoline fire which may have originated from leaks in the fuel lines as the airship had had several fuel leaks in the past, prior to the accident" (Der Bombenkrieg/see also: Schwaebische Zeitung, November 14, 2015) Airship missions to the Balkans were subsequently terminated.

8.2 Timisoara Airship Base Operations After the End of the War

By the end of the war, the door of the airship hangars, barracks for personnel accommodations as well as the warehouses and workshops at Temesvár remained closed (see images 221, 222, and 223).

At the end of WWI, the town of Timisoara and its vicinity (including Szentandrás airship base) experienced a very eventful fate as part of the Banat (The Banat is a historic region in southeastern Europe which is part of present-day Romania, Serbia, and Hungary).

The Banat was initially declared a "Free Republic" with its capital located at Timisoara. It was later occupied by Serbian troops and subsequently "freed" from the Serbs by French troops before it was finally handed over to the Romanians by the French.

The graves of fallen soldiers of various nations are a testament to the fact that this was a time of

Image 224-4 Above: *"Graf Zeppelin II"* airship during "Einhallen" (image: Schmeelke Archive).

bloodshed" (see ADZ, 4/2014 and image 224).

"After the Banat was transferred to Greater Romania, the Sanktandreas-Neubeschenowa airship base was plundered and devastated by Neubeschenowa residents on May 5, 1919. During its disassembly, the airship hangar collapsed." (ADZ, 5/2014)

The airship base including its airship hangar, along with a variety of facilities was strategically established in Szentandrás near Timisoara in 1915. In addition to the existing railway connection and economic ways (Timisoara was among the first European cities to have electrically powered street lighting), its strategic location was of great significance.

The construction of an airship hangar in Szentandrás and the establishment of airship traffic at the Timisoara foothold (including the infrastructure provided for extensive repairs on airships), the city and its surroundings became a domicile for technological innovations such as

- The establishment and operation of a gas works for hydrogen production and storage, including the creation of an artesian well (with a depth of 200 m) with two large water pools for water supply which also served an open-air swimming pool in the summer.
- The establishment and operation of a machinery house for electricity generation as well as the creation and operation of a power line leading to the gas works and other airship base facilities.
- The establishment and operation of radio communication stations including

telegraphy to communicate with airships, determine their position as well as transmit important weather information.

- An electrical workshop for the repair of electrical equipment such as transformers, E-engines, generators, radios, etc. to secure operations at the airship base and various other features onboard the airship.
- An engine workshop for the installation and dismantling of airship engines (internal combustion engines) as well as the repair of important engine accessory equipment to maintain airship flight safety.
- A mechanics workshop for the repair of measuring equipment, transmissions, bearings, etc. as well as the replacement of damaged aluminum frame components on the airship, including the maintenance of the large hangar doors to secure the operation of the airships at the base and their various functions.
- A weather station for the extraction and analysis of weather data as well as the evaluation of winds (wind characteristics) as part of an extensive network to supply airship commanders with important data for flight planning. These weather stations recorded collected data on temperature, wind direction and speed, atmospheric pressure, frontal systems, and cloud cover. By utilizing weather balloons and kites, high altitude atmospheric conditions were measured and recorded.
- A workshop for the maintenance of weather and observation balloons and kites.
- The establishment and operation of a kitchen and a cafeteria for personnel catering (Haltemannschaft, Ballonmannschaft, workshop personnel, etc.) at Timisoara airship base.
- The construction and operation of a railway connection including warehouses for storage of airship base operational equipment.
- The construction and utilization of multiple barracks to accommodate Offiziere and crews.

The above summary of workshops and buildings, equipped in many cases with different technology yields the following questions:

Were these facilities which were equipped with modern technology utilized by craftsmen and businesses in the Banat after 1919?

If the facilities were utilized, how did the use of the still intact equipment affect the further economic development of the Timisoara region?

8.3 Airships – The Tentative End of an Era

"The end of the war marked the tentative end of the Zeppelin airship project which had progressively evolved until 1914.

On June 23, 1919, Luftschiffer followed the 'example' of the Imperial Navy by burning seven of the Zeppelin airships still in existence at Nordholz and Wittmund." (see Bélafi, 1988, page 122)

L 59's journey to Africa in November of 1917 showed that despite issues which arose because of large temperature fluctuations, "it was possible to fly long distances on Zeppelin airships, covering distances which no aircraft would be able to fly without stopping. Important information and data on long-distance airship travel were gathered, crossing a variety of climate zones, which offered a glimmer of hope to LZ GmbH. After the end of the war, the company envisioned a future for its business in long-distance travel." (ZEPPELIN MUSEUM BLOG 2017)

"Many years passed before German airship companies recovered from this setback and started building even larger and better airships, such as LZ 126, LZ 127, and LZ 129. (...)

LZ 127 was obligatorily named '*Graf Zeppelin*'. It gained attention in 1929 when it flew around the world. In 1931, the airship was part of an expedition trip into the eternal ice of the Arctic. The same year, LZ 127 began scheduled operations to North and South America; in 1936, the larger airship LZ 129 followed suit. In that way, both airships contributed to the goal of uniting people by way of aeronautics and airmail." (see: Bélafi, 1988, pages 122/123) Airship "*Graf Zeppelin II*" was an even larger construction (length: 245m / largest diameter: 41.20m / volume: 200.000 m³ / crew: 45 / passengers: 40) (see images 224-2 through 224-4).

9. 2019 – The Search for Traces of Existence of Timisoara Airship Base

In 2019, Jiri Racil, a Czech historian from Brno specializing in WWI events, visited the grounds of the former Timisoara airship base. He was inspired by my article on the Timisoara airship base published in Fliegerrevue X 74 magazine. He searched for fragments of existence of code name "Adebar". Mr. Racil generously provided me with his findings. I attempted to match possible objects captured in images taken by my father or mentioned in the site description with the objects Mr. Racil discovered (see Weber/Petri-1981: SANKTANDRES, pages 164-166).

The objects discovered by Jiri Racil were numbered, their respective GPS data was recorded and entered on a map (see image 225).

Two concreted basins (object 1/see image 226), one of which today is utilized for pig farming, could not be overlooked. These two pools, filled with water from four artesian wells were part of the gas works; they were used to produce hydrogen when the airship base was still in operation. The pools were also utilized by airship base personnel to take a cool dip in the summer (see image 227). During the Romanian era, these two basins were "a very popular swimming spot for the Sanktandres and Neubeschenowa population; people even traveled from Timisoara to use the pools here." (SANKTANDRES/1981, page 169).

The concrete pipe which was discovered (object 2/see image 228) could have been part of the underground line which led from the railway depot to the engine room where the airship's cells were filled in the hangar (SANKTANDRES/1981, page 165). Alternatively, it could have been a pipe utilized to connect the line(s) of the four artesian wells at a depth of 200 meters.

The concreted cellar (object 3/see image 229) could be the object described on page 165 (SANKTANDRES/1981) titled "bomb cellar". "North of the hangar there was a large, fortified cellar where bombs were stored under lock."

The concrete blocks (object 4/see image 230) appear to have been compiled over the years, most probably originating from a variety of places. Based on the slightly offside location, it is assumed that the collection point was established in the location where the future new beacon was going to be built

Image 225 Above: Locations of the discovered objects (Racil Collection)

(at a small distance from the airship hangar, see image 231). The steel structure was anchored to the ground at four points by a concrete foundation (made of concrete blocks).

Such concrete blocks were also utilized as a foundation for combustion engines and generators in the machinery room (see image 232).

The discovered brick foundation (object 5/ see image 233) could have been the base of the machinery house where DC was produced; this building had an extensive brick foundation (see image 234). It is entirely possible that the machinery house was built with a cellar for the storage of fuel

Image 226 Facing Page, Above: Object 1 – Pig farming in the former gas works pool (image: Racil Collection)

Image 227 Facing Page, Below: Swimming in the gas works pool (image: Biber Archive).

Image 228 Right: Object 2 – large concrete pipe (image: Racil Collection)

Image 229 Below: Object 3 – possible remnants of the former bomb cellar (image: Racil Collection)

Image 230 Above: Object 4 – multiple concrete bases (image: Racil Collection)

Image 231 Left: New beacon structure – metal frame anchored in four concrete foundations (image: Biber Archive)

Image 232 Facing Page, Above: Machine room for DC generation (image: Biber Archive).

Image 233 Facing Page, Below: Object 5 – brick foundation (image: Racil Collection)

and combustion engines.

The concrete ramp (object 6/**see image 235**) is located in the immediate vicinity of the pools and the former gas works where gas was produced and filled into gas cylinders. The ramp is flat which could mean it was utilized to transport heavy gas cylinders (for example on trolleys/**see image 236**) onboard the airship. As the grounds in the area surrounding the hangar were swampy, these paths had to be firmly secured.

Image 234 Above: The brick foundation for the machinery house (used for DC power generation) is clearly visible (image: Biber Archive).

Image 235 Left: Object 6 – Concrete ramp/ concreted transport route (image: Racil Collection).

Image 236 Facing Page, Above: Transportation of gas cylinders on trolleys in the gas shed (image: Biber Archive)

Image 237 Facing Page, Below: Object 7 – Vented water pipes (image: Racil Collection).

The valved water pipe (object 7/ see image 237) is no longer in use. Back in the day, the water was transported from four artesian wells at a depth of 200 m to both water basins at the gas works via these long water pipes. The valves and fittings allowed for flow regulation of the water quantity.

These rather old looking objects (see image 238) were possibly lightning conductors. It is assumed that they were used to keep the large airship

Image 238 Above: Object 12 – old structures, possibly used as lightning conductors (image: Racil Collection)

Image 239 Left: Tall, streamlined structures (possibly lightning conductors) located adjacent to the airship hangar (image: Biber Archive).

Image 240 Facing Page, Above: View of Szentandrás / Sánandrea (image: Biber Archive).

Image 241 Facing Page, Below: View of Szentandrás / Sánandrea (image: Racil Collection).

Image 240 Above: View of Szentandrás / Sánandrea (image: Biber Archive).

Image 241 Above: View of Szentandrás / Sánandrea (image: Racil Collection).

hangar from being hit by lightning. It is also possible that they served as masts for the wireless telegraph station as well as for the radiostation. Lastly, there is a possibility that the round steel rods extending from the tip into the ground were a combination of lightning conductors and anntenae masts.

Image 239 shows streamlined structures facing upward in the vicinity of the airship hangar. These objects are assumed to be the lightning conductors mentioned above and pictured in image 238s.

The objects shown appear to be the last remaining relics of the impressive Timisoara airship base with its large airship hangar, many workshops, depots, accommodations, cafeterias, and wards. Unfortunately, a vast number of materials were removed from Timisoara airship base after the unexpected departure of base personnel; furthermore, the lack of utilization of the workshops took a toll on their condition.

The remark in the SANKANDRES/1981 book on page 169 was a clear reflection of the situation: "When the orthodox priest Pavel Jurma was tasked with the administration of the airship hangar inventory during the Romanian era, there was not much left for him to tend to."

Finally, let us take one last look from Timisoara airship base towards Szentandrás / Sánandrea to compare the past (see image 240) and the present (see image 241).

10. Conclusion

In Paul Biber's estate, there were glass slab photographs containing images of balloons in operation; unfortunately, these did not survive the last 60 years. It is assumed that these images originated from his time at Timisoara airship base. They were an indication of just how interested Paul Biber was in the operation of weather and observation balloons. It is possible that Paul Biber was offered an opportunity to partake in exercises with the balloon crews at Timisoara airship base and to ride along in one of the balloon baskets.

Based on his civilian working activity with Firma Rosenmüller, where he manufactured equipment for altitude and speed measurement, he was also involved in the execution of balloonist tasks.

By partaking in one or multiple balloon rides, Paul Biber would have had the opportunity to prove his "flight aptitude" as well as his ability to deal with the effects of high-altitude operations. Paul Biber was specifically selected for his new task. The ability to deal with the effects of high-altitude operations as well as flight aptitude were basic prerequisites for his new role with 1. Seeflieger-Abteilung Holtenau.

After a long journey from Timisoara, former aeronaut (Stabsgefreiter) Paul Biber reported for duty with 1. Seeflieger-Abteilung Holtenau at the end of August 1916. He had to prove his professional competence for the task he was selected for, both in theory and in practice. It appears his results more than satisfied his superiors.

Lastly, Paul Biber had to prove his physical aptitude. In October 1916, he strapped into a Navy aircraft which performed a variety of flight maneuvers. Paul Biber passed his aptitude test on an aircraft with flying colors.

Soon after passing his aptitude test, Paul Biber took over the establishment and leadership of Versuchs-und Reparaturwerkstatt (experimental workshop and repair facility) at Seeflugzeugversuchs-Kommando Warnemünde (see images 242 and 243, stamp of Paul Biber's new duty station Flugverschskommission Warnemündes). He participated in many speed-measurement flights himself (see images 244 and 245). His area of responsibility also included the proper measurement, installation, repair, and improvement of all aeronautical instruments (see Biber 2018). Almost all 2138 Navy aircraft were tested at Seeflugzeugversuchskommando Warnemünde. In March 1918, Paul Biber passed the theoretical exams for Flugmechaniker-Maat in Kiel; he was promoted on May 1.

After the cease-fire on November 11, 1918, Paul Biber left SVK. He returned to work as a foreman at Firma G. Rosenmüller. He was voted chairman of "Verein der Mechaniker und Optiker zu Dresden and Umgebung" and submitted his first patent specification in 1920. He passed his master's certification exams with distinction and, in 1921, Paul Biber was selected to work as "Fachlehrer und Werkstattleiter der Feinmechanischen Abteilung" (specialist subject teacher and workshop manager of the precision and fine mechanical department) at the famous German Glashütte Uhrmacherschule (watchmaking school).

His extensive knowledge and experience were incorporated into *"Taschenbuch für Feinmechanik"* which he co-authored. The paperback book was published in 1930.

On September 1, 1957, Paul Biber passed away in Glashütte. His former precision mechanic and master course students ensured that their teacher and lecturer was honored with a commemorative plaque at Ruhestätte verdienter Männer der Deutschen Uhrmacherschule (resting place of deserving German watchmaking school members) (see image 246). Throughout the exhibit, Deutsches Uhrenmuseum Glashütte (German watchmaking museum) highlighted Paul Biber's accomplishments to secure excellent training of precision mechanics, watchmaking masters, and other students at German Uhrmacherschule Glashütte (see image 247).

On the occasion of my father Paul Biber's 125th birthday in 2016, the German Watch Museum Glashütte presented a special exhibit on the life and work of the Head of the German Watchmaker School Department of Precision Engineering; this temporary exhibit complemented Paul Biber's achievements outlined in the museum's permanent exhibition. As his son, I was particularly pleased to gather and file a large number of images, documents, articles, books, and equipment with the help of staff members from the German Watch Museum Glashütte in order to present a vivid image of my father to an audience of elderly former students as part of a lecture evening (see image 248). My considerations were focused on his work at the German Watchmaker School (Deutsche Uhrmacherschule), including the establishment of the Engineering School for Precision Mechanics

Image 242 Left: Flugmechaniker-Maat at Seeflugzeugversuchskommando Warnemünde (image: Biber Archive).

Image 243 Above: Stamp of Paul Biber's new duty station – "FLUGVERSUCHSKOMMISSION WARNEMÜNDE" – later "SEEFLUGZEUGVERSUCHSKOMMANDO WARNEMÜNDE" (Biber Archive).

Image 244 Below: Paul Biber in December 1916, ready for a speed measurement flight at Seeflugzeugversuchskommando Warnemünde (image: Biber Archive).

Image 245 Above: Hansa-Brandenburg W29 following Paul Biber's measurement flight on July 9, 1918 (image: Biber Archive).

Image 246 Below: Author Dr. Jörg Biber next to the commemorative plaque honoring his father, Paul Biber (image: Dr. Jörg Biber)

Image 247 Above: Teachers, including Paul Biber (third from right) with the student group (image: Dr. Jörg Biber)

Image 248 Below: Lecture about Paul Biber at Deutsches Uhrenmuseum Glashütte on September 22, 2016 (image: Dr. Jörg Biber)

(Ingenieurschule für Feinwerktechnik) and its impact on the watchmaking industry in Glashütte. During my research, I encountered a series of documents outlining his work experience as an aeronaut at Timisoara airship base as well as his role as Head of Experimental Research and Repair Facility at Seeflugzeugversuchskommando Warnemünde. These documents formed the foundation of extensive research which ultimately led to the writing of the "Code name Adebar" article series in *Banater Zeitung* (Banat newspaper), as well as this book.

On behalf of the entire Biber family, I would like to thank you for your interest in my father's experiences in Timisoara.

If you enjoyed my father's experiences with Königlich Sächsischer Feldtrupp für Luftschiffe Nr. 14 and would like to learn more about Paul Biber, please refer to my other book, Paul Biber and the Seeflugzeugversuchskommando Warnemünde 1916–1918.

Paul Biber and the Seeflugzeug Versuchs Kommando Warnemünde 1916–1918 by the same author is also available from Aeronaut Books. This is the author's story of his father, a talented precision mechanic, at Germany's seaplane test command during the last two years of the war. In addition to the author's personal story, the book gives an in depth look into the SVK and its operations that has never been presented before. This unique story is heavily illustrated with 357 photos, many previously unpublished, 45 color profiles, 47 drawings, and dozens of miscellaneous graphics within its 294 pages.

Excerpts from the review in *Cross & Cockade International* Autumn 2021, Volume 52/3:
"The book is illustrated by more than 360 photographs depicting not only Biber and the various aeroplanes mentioned, but also many original advertisments, letters, and similar documents, technical specifications, and period drawings."

"Enthusiasts for German floatplanes will not doubt find this book indispensible and anyone with an interest in the more technical aspects of WWI will no doubt find it as facinating as I did myself."

11. Overviews

11.1. Overview of Airship Operations in the Balkans at Airship Bases in Timisoara and Jamboli

Dates/Timisoara	Timisoara Airship Base	Jamboli Airship Base	Dates/Jamboli
October 28, 1915	Completion and commission of Timisoara airship base		
November 2, 1915	Arrival of LZ 81 at Timisoara (departed from Dresden)		
November 8, 1915	Departure of LZ 81 for Sofia with special ambassador to the German emperor, Herzog Johann Albrecht von Mecklenburg		
November 9, 1915	Arrival of LZ 81 in Sofia with special ambassador to the German emperor, Herzog Johann Albrecht von Mecklenburg. Alliance discussion with Bulgarian Zar Ferdinand Departure of LZ 81 for Timisoara		
November 10, 1915	Arrival of LZ 81 in Timisoara		
November/December	LZ 81 training trips to practice Aus- and Einhallen (moving the airship into and out of the hangar) as well as Steigen and Fallen (raise and drop)		
December 31,1915 and January 1, 1916	Planned war patrol flight of LZ 81 to Saloniki/cancelled due to engine damage		
January 19, 1916	Aushallen of LZ 81 and return to Dresden		
January 27, 1916	Arrival of LZ 85 in Timisoara from Allenstein		
January 31 and February 1,1916	War patrol flight (Hptm. Scherzer) of LZ 85 to Saloniki/Drop 2000 kgs of bombs		
February 8, 1916	LZ 85 war patrol flight to Saloniki/cancellation due to poor weather near Nis		

Dates/Timisoara	Timisoara Airship Base	Jamboli Airship Base	Dates/Jamboli
March 17, 1916, through March 19, 1916	LZ 85 war patrol flight to Saloniki/drop of 2000 kgs of bombs/ stopover on March 18 after hitting targets in Sofia/ onward journey (Hptm. Scherzer) with 11 remaining crew members onboard and missing starboard engine/ arrival in Timisoara on March 19		
Late March, 1916	Starboard engine and remaining crew members arrive at Timisoara from Sofia by railway		
Late March through late April	As a result of extensive repair works by LT 14 maintenance crews, LZ 85 regains its flight capability		
May 4, 1916, through May 5, 1916	LZ 85 war patrol flight to Saloniki/ drop of 2000 kgs of bombs/ heavy hits forced LZ 85 to make an emergency landing in the Varda swamps/ the crew destroyed LZ 85/ Hptm. Scherzer and the crew of LZ 85 are captured		
		Completion and commission of Jamboli Airship Base, approximately 80 km west of Burgas	Mid-June
June 21, 1916	SL 10 stopover at Timisoara enroute to Jamboli airship base		
June 23, 1916	SL 10 ascends from Timisoara airship base and onward journey to Jamboli airship bases	SL 10 landing at Jamboli airship base	June 24, 1916
		SL 10 carries out many reconnaissance flights across the Black Sea/ 16-hour reconnaissance flight to Zonguldak on the southern coast of the Black Sea (Hptm. Wobeser)	Late June through late July July 15, 1916

Dates/Timisoara	Timisoara Airship Base	Jamboli Airship Base	Dates/Jamboli
June and July 1916	LZ 86's length is extended in Dresden in preparation for its transfer to Timisoara where it was to replace LZ 85	During a war patrol flight to Sewastopol, SL 10 went missing overhead the Black Sea (it is assumed that LS 10 crashed following a lightning strike)	July 28, 1916
		SL 10 was immediately replaced by LZ 101/ LZ 101 departs for Jamboli/ LZ 101 was caught in a heavy storm and made a stopover landing at Timisoara airship base	July 31, 1916
August 1 through August 3, 1916	LZ 101 was repaired in Timisoara by LT 14 maintenance crews	LZ 101 landed at Jamboli	August 3, 1916
		LZ 101 attacked Bucharest/ dropped 1500 kg worth of bombs and one large bomb (total of 2000 kg in bombs; Hptm. Gaissert)	Night of August 28 through 29, 1916
August 30, 1916	The converted airship, LZ 86 flew from Dresden to Timisoara via Vienna/LZ 86 arrived at Timisoara airship base		
September 3, 1916	LZ 86 first war patrol flight/ attack on Bucharest und Ploiesti/Drop of 2000 kg worth of bombs		
Midday on September 4, 1916	LZ 86 reached Timisoara unscathed when it suddenly dropped; one of its nacelles was destroyed on impact, the second nacelle was destroyed when the airship crashed/LZ 86 was no longer operational and was dismantled/ a memorial service was held in Timsoara for the crew members who perished in the crash (including Hptm. Walter Wolff).	LZ 101 attacks Bucharest at night, dropping 2000 kg worth of bombs (Hptm. Gaissert)	September 4–5, 1916
Mid-September	LZ 86 was replaced by LZ 81, an extended airship version/LZ 81 departed Dresden for Timisoara via Vienna, under the command of Hauptmann Barth.		

Dates/Timisoara	Timisoara Airship Base	Jamboli Airship Base	Dates/Jamboli
September 24–25, 1916	LZ 81 departed Timisoara to launch an attack on Bucharest. The city of Bucharest was hit by many bombs. The following day, LZ 81 reached its home base of Timisoara almost unharmed.		
September 26, 1916	LZ 81 departed Timisoara on its second war patrol flight at 17:00, dropping 1650 kg of high explosive and incendiary bombs on Bucharest (Hptm. Barth).	LZ 101 dropped 2000 kgs of bombs on Bucharest departing from Jamboli (Hptm. Gaissert).	September 27, 1916
September 27, 1916	During both nights, LZ 81 dropped a total of 4513 kg worth of bombs on Bucharest (Hptm. Barth). LZ 81 received multiple severe hits overhead Bucharest. On the return flight, the airship experiences a hard landing near Tirnova (Bulgaria). Two crew members were injured, one of which later succumbed to his injuries. Repairing LZ 81 was not worth the effort- airship LZ 81 was subsequently dismantled.	O.H.L. bomber wing 1 aircraft (Kampfgeschwader 1) also dropped bombs on Bucharest.	
Early October October 17, 1916	Kommandant Helmut Weidling received orders to fly to Timisoara from his home base of Namur on LZ 97/ LZ 97 arrived at Timisoara airship base	LZ 101 departed on a war patrol flight to Ciulnita near Calarasi, carrying approximately 1600 kg of bombs (Olt. Koereuber).	October 5/6, 1916
October 23 through October 24, 1916	LZ 97 (Timisoara airship base/Olt. Weidling) and LZ 101 (Jamboli airship base Olt. Koreuber) depart on a joint war patrol flight to bomb Bucharest.		October 23/24, 1916
October 25/26, 1916	LZ 97 war patrol flight (Timisoara airship base/ Olt. Weidling) and LZ 101 (Jamboli airship base/Olt. Koreuber) dropped bombs on railway junctions in Ciulnita, Fetesti, and Ploesti. Bucharest was conquered by German army units on December 6, 1916		October 25/26, 1916
		LZ 101 successfully carried out war patrol flights to Galatz, Mudros, and Odessa (among others)	November/ December 1916

Dates/Timisoara	Timisoara Airship Base	Jamboli Airship Base	Dates/Jamboli
s		LZ 101 attacked Galatz in Romania a second time (Olt. Koreuber), dropping 12 explosive bombs weighing 58 kg each	December 25/26, 1916
Late December 1916	Romania was occupied by German army units		Late December 1916
January through April 1917	As there were no further targets for LZ 97 in Romania, LZ 97 carried out war patrol flights on multiple Italian ports (Valona, Brindisi, and Taranto).	LZ 101 carried out its last successful attack on Mudros (Lemnos), carrying approximately 1600 kg of bombs. This was the very last bombing mission carried out by an Army airship.	March 20/21, 1917
April 25, 1917	The success of the distant reconnaissance flights was hampered by poor weather. From a navigational perspective, these flights were noteworthy. The last war patrol flight by LZ 97 on Valona was cancelled due to bad weather	LZ 101 was to attack English footholds on the Aegean Islands and in Odessa. LZ 101's attempts to attack Odessa, Mytiöene, and Jassy failed.	Summer of 1917
May 5, 1917 July 5, 1917	Timisoara airship base was abandoned/shut down LZ 97 departed Timisoara for Jüterbog (via Vienna), where it was decommissioned and disarmed	Jamboli airship base was taken over by the Navy LZ 101 (one of the most successful Army airships/ dropped 11,934 kgs worth of bombs) departed Jamboli.	September 1917
		LZ 101 was transferred from Jamboli to Schneidemühl. LZ 101 was decommissioned and disarmed.	August 31, 1917
		Present-day Jamboli Naval airship base received L 59, a new Navy airship (=LZ 104, equipped with the largest airship shell with a volume of 68,500 m³ and a length of 226.5 m)	
		L 59 deaprted on a long-haul flight to Africa (Supply of General Letow-Vorbecks's troops)	November 21, 1917

Dates/Timisoara	Timisoara Airship Base	Jamboli Airship Base	Dates/Jamboli
s		L 59 reached the African coast near Ras Bulau	November 22, 1917
		L 59 received orders to turn around approximately abeam Kharttum	November 23, 1917
		L 59 landed at Jamboli, its crew completely overtired after a 95-hour journey. L 59 traveled 6,757 km in 95 hours, a new Zeppelin airship track record.	November 25, 1917
		L 59 flew from Jamboli to Germany where it was converted into a reconnaissance and attack airship.	December 1917
		L 59 carries out valuable reconnaissance flights from Jamboli across the Mediterranean Sea and two war patrol flights on Naples and Malta. While on the war patrol flight to Malta, L 59 crashed into the Mediterranean burning overhead the Strait of Otranto (it is assumed that the airship crashed due to a gasoline fire)	February to April 1918
	Termination of airship missions in the Balkans including the Black Sea and Mediterranean from Jamboli and Timisoara airship base		

11.2 Airships Operating in the Balkans

LZ 81: (original): Length = 163.5 m
Army airship LZ 81 (LZ 51), p construction series, volume = 32.000m³, largest diameter = 18.70m, crew = 19, stationed at Timisoara airship base from November 2, 1915 – January 19, 1916

LZ 85: Length = 163.5 m
Army Airship LZ 85 (LZ 55), p construction series, volume = 32.000m³, largest diameter = 18.70m, crew = 19, stationed at Timisoara airship base from January 27, 1916 – May 4, 1916; stranded in the Vardar swamps on May 5, 1916 and subsequently destroyed by its crew

SL 10: Length = 173.95 m
Army Airship Schütte-Lanz SL 10, e construction series, volume 38.000m³, largest diameter = 20.10m, crew = 12-14, stationed at Timisoara airship base from June 24, 1916 – July 28, 1916; possibly crashed over the Black Sea after being struck by lightning on July 28, 1916

LZ 86: Length = 163. 5m + 15.0m = 178.5m (extended in Dresden by 15m)
Army Airship Zeppelin LZ 86 (LZ 56), p construction series, volume increased to 35.800m³ in Dresden, largest diameter = 18.7m, crew = 19. LZ 86 impacted the ground upon safely returning to Timisoara on Augusst 30 – September 4, 1916; both nacelles were destroyed on impact. LZ 86 was subsequently disarmed.

LZ 81: (modified): Length = 163.5m + 15.0m = 178.5 m (extended in Dresden by 15m)
Army Airship Zeppelin LZ 81 (LZ 51), p construction series, volume increased to 35.8000m³ in Dresden, largest diameter = 18.7 m, crew = 19; stationed at Timisoara airship base from September 15 – 27, 1916, hard landing near Tirnova (Bulgaria) following bomb drop on Bucharest; LZ 81 was subsequently disarmed.

LZ 97: Length = 178.5 m
Army Airship Zeppelin LZ 97 (LZ 67), q construction series, volume = 35.800m³, largest diameter = 18.7m, crew = 19, stationed at Timisoara airship base from October 17, 1916 – July 5, 1917; LZ 97 was subsequently disarmed at Jüterborg.

LZ 101: Length= 178.5m
Army Airship Zeppelin LZ 101 (LZ 71), q construction series, volume = 35.800m³, largest diameter = 18.7m, crew = 19, stationed at Jamboli airship base from August 3, 1916 – September 1917; LZ 101 was subsequently disarmed at Schneidemühl.

L 59: Length=226.5m
Navy Airship Zeppelin L 59 (LZ 104), w construction series, volume = 68.500m³, largest diameter = 23.9m, crew = 26, stationed at Jamboli airship base from November 1917 – April 7, 1918; long distance flight from Jamboli to abeam Kharttum (Africa) and back to Jamboli from November 21-25, 1917, new track record of 6.757 km traveled in 95 hours; L 59 crashed burning overhead the Strait of Otranto on April 7, 1918 (possibly due to a fuel fire).

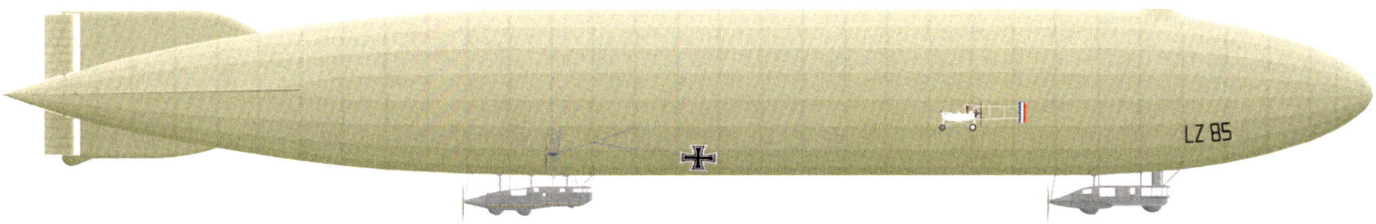

LZ 85 with Voisin 3 for size comparison.

11.3 Overview of "LT 14" (Royal Saxon Field Forces for Airships Number 14) Deployment Sites

When Germany joined the war on August 1, 1914, the October 9, 1913 mobilization schedule came into effect. In this context, airships which had been utilized for civilian operations were seized for military use. To guarantee a high degree of operational readiness, multiple airship forces were established. In Dresden, the Royal Saxon Field Forces for Airships Number 14 ("LT 14") was established. The book talks about multiple "LT 14" deployment sites, highlighting the troop's geographic locations. The most significant deployment sites are presented in the below overview.

Militär-Luftschiffhafen Dresden-Kaditz (Military Airship Base Dresden-Kaditz)

August 1, 1914, through February 14, 1915 (see Schneevogt/Schulz 2007, page 33)

As of August 1, 1914, the military took over the Dresden airship base including all its installations, renaming it "Dresden Military Airship Base" (see Schneevogt/Schulz 2007, page 30).

By the outbreak of the war, 3. Sächsische Luftschiffer-Kompanie (3rd Saxon Aeronaut Kompanie) was converted to Royal Saxon Field Forces for Airships number 14 (also referred to as airship troop number 14; see Schneevogt/Schulz 2007, page 9). "LT 14" immediately took over security operations at Dresden-Kaditz military airship base. Hauptmann G. v. Milczewski was the troop commander (see Schneevogt/Schulz, page 32).

Paul Biber was a member of "LT 14" from December 31, 1914.

When the Naval Forces took over the Dresden airship hangars on March 15, 1915 for training purposes, Royal Saxon Field Forces for Airships Number 14 ("LT 14") departed Dresden for Schneidemühl airship base on February 14, 1915 (see Schneevogt/Schulz 2007, page 32).

Schneidemühl Airship Base

February 15, 1915, through May 10, 1915 (see Schneevogt/Schulz 2007, page 33)

Royal Saxon Field Forces for Airships Number 14 took on several duties to secure operational functionality of the airship base. The large airship hangar which could accommodate two airships, was situated approximately 3 km north of the city of Schneidemühl.

"On May 10, 1915, LT 14 Troop Commander, Ltn. Helmuth Weidling, reported its relocation to Jüterbog to the Royal Saxon War Department." (Schneevogt/Schulz 2007, page 32)

Jüterbog Airship Base

May 11, 1915, through October 16, 1915 (see Schneevogt/Schulz 2007, page 33)

In May, Paul Biber transferred to Jüterbog airship base along with "LT 14". The troop was housed at Dennewitz, where Paul Biber stayed at the home of civilians in a Dennewitz neighborhood. During their deployment to Dennewitz, members of LT 14 were utilized to carry out construction work for the airship base.

Army airship Z IV (civilian LZ 16) landed at Jüterbog airship base on September 13, 1915. Training began immediately. Utilizing airship Z IV, LT 14 members practiced essential operations for Aushallen (moving the airship out of the hangar) and Aufstieg (airship ascent), as well and Landen (landing) and Einhallen (moving the airship into the hangar).

From May through October 1915, Paul Biber attended aeronaut training at Jüterbog airship base. It was integrated into his duties as a member of Landungstrupp.

While at Jüterbog, Paul Biber had at least one opportunity to participate in a trip onboard airship Z IV.

In late October, the newly established Field Forces for Airships Number 22 took over the support of training airship Z IV (see Schneevogt/Schulz 2007, page 34).

Timisoara Airship Base (Southern Hungary – present day Romania)

October 17, 1915, through February 1, 1917 (see Schneevogt/Schulz 2007, page 33)

Following orders issued on October 10, 1915, "LT 14" was transferred to Timisoara. "On October 16, 1915, LT 14 troop commander Oblt. Deubel, transfer to Timisoara in accordance with Feldflugchef orders." (Schneevogt/Schulz 2007, page 32)

Transportation of LT 14 troop members as well as vehicles and materials necessary for the operation of Timisoara airship base was accomplished via railway.

Paul Biber arrived Timisoara airship base via railway on October 22, 1915.

Upon completion of the airship hangar and set up of all workshops and warehouses for the operation of the airship base, airship LZ 81 landed at Timisoara on November 2, 1915, supported by LT 14.

A variety of airships (LZ 81, LZ 85, LZ 86, LZ 97) operated in the Balkans region from Timisoara airship base in succession.

Paul Biber left Royal Saxon Field Forces for Airships Number 14 on August 31, 1916, transferring to Seeflugzeugversuchskommando Warnemünde.

Field Forces for Airships Number 23 was stationed at Timisoara airship base for a short time. LT 23 was solely responsible for the security of Timisoara airship base operations between February and early May 1917, when LT 23 troops were withdrawn from Timisoara.

When Armed Forces airship operations ended in early May 1917, Timisoara airship base was terminated as an operational base for airships. On May 5, 1917, Army airship LZ 97 departed Timisoara for Jüterbog via Vienna. It was scrapped upon arrival at Jüterbog.

Wittmundhafen Airship Base

August through November 1916 (see Schneevogt/Schulz 2007, page 33)

Parts of Royal Saxon Field Forces for Airships Number 14 were relocated from Timisoara airship base to Wittmund (Aurich district near Bremerhaven) in preparation for the airship base's planned operations with airship LZ 90. The troop members assisted in construction work for the new airship hangar (which could accommodate two airships) as well as important facilities for the operation of the base.

In early 1917, Navy airship troop number 4 took over operations at Wittmundhafen airship base. The Royal Navy transferred three airships there (see Schneevogt/Schulz 2007, page 33 and NDR.de, November 4, 2016).

It is likely that all LT 14 personnel which had temporarily been relocated to Wittmundhafen returned to Timisoara upon completion of their tasks.

Jamboli Airship Base (Bulgaria)

February 2, 1917, through May 31, 1917 (see Schneevogt/Schulz 2007, page 33).

"Message from LT 14 addressed to acting Gen.-Komm. XII: as of February 1, 1917, troop has been relocated to Jambol (old spelling of Jamboli) in Bulgaria following inspection by aeronaut troops." (Schneevogt/Schulz 2007, page 32).

Personnel numbers there were increased by members of LT 14 as a large number of aeronaut troop number 18 soldiers stationed at Jamboli were suffering from dysentery and malaria.

Having LT 14 members at Jamboli assured the operational readiness of airship LZ101. Personnel support was particularly important during the period from February through late May/early June of 1917.

The departure of airship LZ101 from Jamboli airship base on August 31, 1917, for Schneidemühl marked the end of Army airship operations at Jamboli. LZ 101 was scrapped after arriving at Schneidemühl.

Thereafter, Jamboli airship base was taken over by the Navy; in November 1917, Navy airship L 59 (LZ 104) arrived at the base.

Weiterstadt Airship Base (Darmstadt)

By order of the commanding general of the Air Force dated May 19, 1917, Royal Saxon Field Forces for airships were dissolved; personnel were transferred to Ersatz-Abteilung des Luftschiffer Bataillon Nr. 2 in Darmstadt, where they subsequently served as a Saxon Detachment (see Schneevogt/Schulz 2007, page 33).

According to a different source, the deployment of LT 14 did not take place at Weiterstadt airship base until March of 1918 (see Stadtwiki Dresden). This account coincides with the following statement: "Mail receipts and documents continue to surface bearing the stamp of LST 14 (LST 14 and LT 14 are both commonly used; author's note) from Weiterstadt near Darmstadt dated from the summers of 1917 and 1918. At this location, LEA number 2 staff members formerly based in Berlin were stationed starting in 1917; their duties included working for Fesselballon units. Saxon personnel were added to the unit as well." (Schneevogt/Schulz 2007, pages 35/36)

12. History Is Alive

12.1 Interview with Author Dr. Jörg Biber, conducted by Raluca Nelepcu

"The accounts characterize an intact world in the vicinity of Timisoara. War appears to be far away."

On July 2, 1900, when the very first airship successfully ascended into the sky above Lake Constance, the airship era had officially begun. A few years later, airships were utilized in World War I. During the war, there was an airship base near the town of Temeswar/Timisoara in the county of present-day Sanktandreas/Sánandrei (formerly Szentandrás). Over the course of time, traces of the airship base have disappeared almost completely. However, its history could be re-traced with the help of precision mechanic Paul Biber's extensive photographic and written legacy. Paul Biber was stationed at Szentandrás airship base in 1915 and 1916. The photographs and written accounts were inherited by his son, Dr. Jörg Biber who published them in a series of articles as part of Banat newspaper approximately two years ago. In this interview, Dr. Jörg Biber (age 70) shares special memories of his father along with some of his most prized inherited possessions. He also explains why the historic written and photographic legacy of Paul Biber are so important.

1. Precision engineer Paul Biber's experiences as an aeronaut during WWI are expected to be published in a book this year. They were partially published in a series in "*Banater Zeitung*" a few years ago. Why was it important to you to record your father's expriences for posterity and make them available in multiple languages?

This is correct. A total of 22 articles on the subject of my father, Paul Biber's experiences as an aeronaut at Szentandrás airship base near Timisoara (code name "Adebar", 1915–1916) were published in "*Allgemeine Deutsche Zeitung für Rumänien*". Both Werner Kemm and Siegfried Thiel assessed the information contained in the articles as interesting contributions across the generations for the Banat region.

What makes my father's account particularly interesting is that his images and records differ greatly from the usual style of war reporting. The snapshots gathered in his photo albums along with the short descriptions of various events and topics of his military service experiences include leisure activities, which provide a detailed insight into life on an airship base in WWI. For example, he talks about the establishment of Timisoara airship base, naming various objects forming part of the entire installation. His images greatly enhance the documentations of his experiences. Paul Biber also captured various groups of people from Timisoara and the vicinity, describing their activities. In context, he outlined essential activities and tasks carried out by ground personnel during Ein- and Aushallen of airships (moving the airship into and out of the hangar). Accounts of encounters with local farmers are also part of the narrative. The establishment of flower beds in the grounds surrounding the airship hangar is well documented. Images captured with "Russians" (most likely prisoners of war) indicate their involvement in works on the airship base and their apparently humane treatment. Reports on airship war patrol flights appear only secondary to his accounts on everyday life. Paul Biber's experiences encompass a broad spectrum of activities and offer insight into life in the Banat region during WWI, particularly highlighting Szentandrás airship base near Timisoara as well as its surrounding areas. The multifarious perspective in my father's narrative were recognized by Werner Kremm and Professor Rudolf Gräf as suitable historic documents; they worked hard to have this book published in German and Romanian in the Banat region. I am pleased that Raluca Nelepcu, an enthusiastic editor, has taken on the translation of the book as well as its layout. An English edition of the book is planned as well.

2. How many physical items are contained in your father's historically valuable legacy?

There is a photo album with images captured by aeronaut Paul Biber from his experiences during training and his deployment with Königlich-Sächsischer Feldtrupp für Luftschiffe Nr. 14 (LT 14) from 1914–1916. It contains almost 100 photographs and documents (i.e. his military passport) as well as 20 field postcards or images Paul Biber took himself to use them as field postcards. He owned a large plate camera which is clearly discernable in the image taken at the arrival of LT 14 at the Kistelek railway station. Photography enthusiast Paul Biber used his camera to capture many interesting encounters during his deployment in Timisoara. Since owning such an expensive camera during WWI was uncommon, it is assumed that many images documenting the life of LT 14 personnel were taken with my father's camera. It was quite rare for everyday life to be captured in photographs during WWI; we are fortunate to have such a vast number of personal images and accounts to share with the public after 100 years. These documents, along with the many notes and written accounts, form a remarkable historic document.

3. When did you first encounter these documents and when did you consider publishing them?

My father was a lecturer at Ingenieurschule für Feinwerktechnik (school of engineering and precision mechanics) in Glashütte which was part of the well-known Deutsche Uhrmacherschule (German watchmaking school, DUS) Glashütte. My father passed away suddenly on September 1, 1957, two days before my first day of school. My mother left many items used by my father for work at the educational institution since 1921 at the engineering school. She initially saved his private belongings before passing them on to me. When I was a boy, I would occasionally open one of my father's photo albums as his images of airship and naval aircraft fascinated me. I would frequently look at items he made himself, such as the "Würfel im Würfel" (die within a die) or "Zwei ineinandergreifende Zahnräder" (two interlocking notched wheels).

On the occasion of my father's 125th birthday, Deutsches Uhrenmuseum Glashütte informed our family of their intent to host a small temporary exhibit in July of 2016 along with a lecture evening for former engineering school students in September. We were pleased to support their project by preparing items and documents for the exhibit. I also agreed to work on a comprehensive contribution concerning Paul Biber's life and career.

The foundation of my task consisted of many filed documents and items of my father's. This is how I once again encountered the images and accounts of my father from his aeronaut training with Königlich-Sächsischer Trupp für Luftschiffe LT 14 as well as his deployment to Timisoara airship base and his work as head of experimental workshop and repair facility at Seeflugzeugversuchskommando Warnemünde.

While analyzing my extensive and moving lecture about my father's multi-faceted life I came to the conclusion that I would direct my focus to his military fields of activity. Internet research yielded many useful pieces of information which I combined with the many photographs and notes of my fathers to form two extensive articles which I sent to Uwe Jack (editor for *Fliegerrevue X* magazine) in the spring of 2018.

The articles contained a variety of new, previously unpublished images and accounts; he was fascinated and published the articles in the renowned aviation magazine. As only part of my materials fit in the articles, Uwe Jack encouraged me to write two books on the subjects. As a result of the articles' publication, I was contacted by experts who supported my undertaking. Initially, I focused my time and efforts on the subject of Timisoara airship base. Expert Fregattenkapitän a.D. Ulrich Israel supplied me with useful information after peer reviewing my articles.

During my research I also encountered a series of short articles on Timisoara airship base in Allgemeine Deutsche Zeitung (ADZ) für Rumänien newspaper. In the meantime, my collection of materials on the subject had grown substantially. In early 2019, I contacted Werner Kremm, the editor of the newspaper.

This was a lucky choice for me in many aspects. Werner Kremm is competent and enthusiastic; he has many connections, is well-respected with a distinctive social competence. We had a great connection from the beginning, exchanging documents and information via email, involving editor-in-chief Siegfried Thiel in order to determine the 2019 publication dates for the article series which included 19 articles by September. Werner Kremm shared the series with his longtime friend Professor Rudolf Gräf (Klausenburg University prorector until March 2020). Professor Gräf enjoyed the articles and asked to read all of the documents. After reviewing the information, he told me that he would like to publish my manuscript in German and Romanian. Meanwhile I was contacted by historian Jiri Racil from Brno; Uwe Jack had put him in contact with me. Inspired by my article series in *FliegerRevue* magazine, Jiri Racil traveled to Timisoara to search for traces of existence of the former Timisoara airship base. He photographed objects and recorded their locations as JPS data.

He shared his findings with me and together we wrote three supplementary articles for ADZ newspaper in which we analyzed the discovered fragments, assigning them to possible objects from 1915–1918. The expanded information was added to the book manuscript.

4. Paul Biber captured war reports on his field

postcards. What emotions did you experience when reading them? How do you feel about war?

The most interesting part about my father's reports of his experiences is that he hardly talks about Zeppelin war patrol flights in his accounts. His focus was on everyday life, including encounters with people from a variety of environments. He also described his surroundings, for example the typically broad streets with a large ditch in the center and footpaths made from bricks on either side of the street he saw while on a stopover in Kistelek. He was fascinated by the farms with their walls cleanly painted beneath the roof overhang. When he arrived in Timisoara, he described the location of the buildings on the airship base and listed their uses. He referenced the swampy grounds surrounding the airship base, photographed the "port" in the Puszta (Hungarian steppe), bemoaning the fact that there were many melons still in the field between the corn which had not been harvested and were now rotting on the inside.

These accounts provide a small insight into the life on the airship base and its surrounding areas near Szentandrás. They reveal that a variety of technology was available for the operation of the base which had to be maintained, a task which required a large number of ground personnel, especially for Ein- and Aushallen (moving the airship into and out of the hangar). Aeronauts and farmers appeared to have a friendly relationship. Furthermore, it provides an insight into Paul Biber's relationship to his parents, as they exchanged packages in addition to postcards.

His descriptions characterize a seemingly intact world in Timisoara. The war seemed far away. The devastation caused by the bombs dropped from airships in war zones (for example Saloniki and Bucharest) was hardly noticeable by airship crews and ground personnel. Strengthening air defense made it increasingly more difficult for airships to complete their war patrol flights unharmed. There are reports of damage and losses of airships associated with dead or captured crew members. Death was a part of every war patrol flight.

The areas surrounding Timisoara still appeared to resemble an intact world. Crew members who died were recognized with a funeral procession and dignified burials at Timisoara cemetery. The local population attended funerals with great compassion and grave plots continue to be cared for today.

In coordination with Professor Rudolf Gräf I tried to incorporate my father's photographs and accounts into the acts of war in context with the various war patrol flights carried out by airships over the Balkans, in hopes of providing as much of a comprehensive insight of the respective situation as possible.

The theater of war extending from the Balkans to the Black Sea showed that each war party tried to involve new allies in an attempt to gain an edge in the war. Conflict resolution based on negotiations had long been abandoned.

Unfortunately, the war turned plans to explore the civilian utilization of airships into the selective bombing of targets from the air. It was not long before the technical advantages of airships were depleted and airships themselves became targets of military operations. It became increasingly more dangerous to operate airships over enemy territories.

By the end of the war, all facilities at Timisoara airship base, such as equipment for power generation, workshops for engine maintenance and repair of Duralumin Zeppelin segments, the gas works and water basins, railway connection, radiotelegraphy, the artesian wells with their water-pipe system, stable accommodations, the large airship hangar, and many more facilities were left in fully working condition. Sadly, none of the technologically advanced facilities were utilized by the population of the Banat region. The airship hangar was destroyed, and everyone helped themselves to existing items and equipment. Shortly thereafter, there was nothing much left to manage by the community, a chance for economic upturn for the Timisoara region had been wasted.

I am part of the generation of people who were born after WWII, which I am very grateful for. My parents as well as my three older siblings (born in 1936, 1937, and 1941) have experienced the war, especially the final phase as well as the post-war period which was characterized by famine, and they have shared their experiences with me. They witnessed the burning of Dresden from Glashütte in February 1945 and saw the first arriving trains transporting refugees from the bombed city of Dresden. My mother worked as a nurse from the first day after the bombing; she witnessed the tremendous destruction and suffering, trying to care for many wounded. Immediately after the end of the war, my family watched the bombing and evacuation of their hometown Glashütte. They spent their days in their garden shed at the edge of the woods, watching low flying aircraft shoot at soldiers in the woods.

When peace was restored and the situation improved, I was born in 1951, a post-war child. My middle name is "Friedewald"; it was very consciously chosen as it means, "there is finally peace in the woods again".

Any German who visited the Museum near

Dubrovnik showcasing the break-up of former Yugoslavia and the associated Balkan War, especially the Croatian War of Independence should be very grateful that we have averted the 1989/90 military confrontation. Such a conflict would have been disastrous considering the German military capabilities at the time. Politicians of the former GDR and Soviet Union responded level-headedly at the time.

5. Which of your father's experiences made a particularly strong impression on you?

I was impressed by the short-term nature and pace with which the Timisoara airship base was established (decision to establish the airship base: July 29, 1915/start of construction: September 3, 1915), including the construction of the airship hangar measuring 180 meters in length, 30 meters in height, and 30 meters in width, including two large doors (start of hangar segment erection: October 4, 1915), the railway connection, as well as other important workshops and warehouses under challenging circumstances at the time. Airship LZ 81 was able to move into the hangar on November 2.

The photograph captured of soldiers waiting for food distribution gives the impression that German, Austria-Hungarian, as well as Russian (captured) soldiers were treated harmoniously and equally.

I was impressed with the fact that my father was tasked with the responsibility of operating the engine room including the transformer station utilized in the production of power and the transfer of direct current. He was a precision engineer and was selected for this task based on his knowledge and additional qualifications gained during evening school courses. He was put in charge establishing the transmission line to the gas works including the erection of transmission masts. My father was not only a precision engineer, but he also became an acknowledged "electrician".

6. What do you feel makes your father's experiences unique and worth reading?

My father wrote his experiences in a very natural way, his images captured a variety of events which took place on the grounds of Timisoara airship base as well as the surrounding area; his accompanying accounts were without a direct connection to war events. Please also reference my comments to 1. and 4. above.

7. You lost your father when you were barely six years old. What memories do you have of him?

I am certain that some experiences I had with my father were anchored in my memory as a result of his sudden, unexpected passing. I vividly remember the last hours I spent with him before he died. We spent time in our garden at the edge of the woods, clearing away storm damage from our property which included part of the woods. There was a lot of work to be done, yet he took the time to saw off a strong branch and put it across a log on the ground. He sat me on one side of the branch and walked to the other side, trying to use the branch like a seesaw.

I frequently picked my father up from work at school for lunch as we did not live too far from there. In front of his workshop was a big room with a lot of measuring equipment, including electronic measuring devices which I was allowed to switch on and off from time to time. Some of them would print out graphs or numbers which I found to be fascinating. There were three photographs displayed in a brass casing; one of them was a picture of an aircraft, the other two were images of my father. The same photographs are hanging on the wall at my house today. The die within a die as well as the interlocking cogwheels structures were on my father's desk along with a colorful, round paper weight made from glass which had a flower inside of it. I have always been fond of these items which today are in my living room.

On May 1, 1957, my father took me to the "Vogelwiese" (grasslands); I got to have a ride on the merry-go-round. There was also a small toy car, slightly larger than a pedal car, which caught my attention. I got to sit in it and steer it; it went slowly, I did not need to pedal, my father ran alongside the car. I will always remember this experience.

In the summer of 1957, my father visited multiple places in the Republic with my mother and me. We spent time on the Baltic visiting a former student of my fathers in the town of Göhren, where he worked as a watchmaker. From Warnemünde we took a ferry to Hohe Düne where he was deployed from 1916-1918 with Seeflugzeugversuchskommando.

8. Years later, you followed in your father's footsteps by choosing a "metalwork profession". How did you choose this career?

Referring to my answers to question 3, I can add the following; when I was a teenager, I asked my family about my father's job-related tasks. Upon graduating from elementary school, I passed my entry exams to Kinder- und Jugendsportschule (children and youth sports school) in Dresden. I completed grades 9 and 10 at 33. Polytechnische Oberschule (polytechnic secondary school) in Dresden. During school breaks, I completed internships in metal technology and decided (perhaps with my father's legacy in mind) to pursue training in tool making. My education at Betriebsberufsschule (BBS, vocational school) was very good, I learned a lot both professionally and socially. I had excellent instructors and masters and enjoyed learning with my classmates. We still meet regularly. I am told my father was a renowned specialist and an excellent educator. Encouraged by my training, I aimed to emulate my father in becoming a vocational teacher. In order to achieve my goal, I needed an Abitur (school leaving examination, general qualification for university entrance). I decided to take evening classes during my training in order to get my Abitur.

Based on my good grades and my desire to become a vocational teacher, I was utilized as a trainer in "Feinmechaniker mit Abitur" (precision engineer with Abitur) classes while still in training myself. Following 18-months of basic military service, I began my studies at Technische Universität Dresden in September 1973. In 1977, I graduated with a degree in engineering education for mechanical engineering (Diplom-Ingenieurpädagoge für Maschinenwesen). Thereafter, I worked as a vocational teacher at BBS Pentacon. After 6 months, I was offered a lectureship at TU Dresden. In September 1978, I became a scientific assistant at TU Dresden and was involved in the training of vocational teachers for mechanical engineering. In addition, I taught at BBS Pentacon. In August 1982, I did my doctorate at TU Dresden, became a university professor. I got involved with teacher training and in 1987, I was appointed senior teacher (Oberlehrer) by Ministerium für Hoch- und Fachschulwesen (Ministry of Higher Education and Academy of Sciences of the GDR), just like my father. I diligently published (more than 250 publications), was a member of professional boards (Berufsfachkommission Werkzeugmacher, Beiratsmitglied der Zeitschrift Berufsbildung), and was involved in multiple BIBB federal model tests (Bundesmodellversuche). I trained vocational teachers for metal and mechanical engineering at Technische Universität Dresden for more than 40 years.

After the "Wende" (change from the rule of the Socialist Unity Party of Germany and a centrally planned economy to the revival of parliamentary democracy and market economy in the German Democratic Republic), my wife and I were invited to annual meetings of former Deutsche Uhrmacherschule Glashütte students. During the meetings I realized how highly esteemed my father was by his students, both socially and professionally. It was his students who dedicated an honorary plaque for my father at the "Ruhestätte verdienter Männer der Deutschen Uhrmacherschule Glashütte". In retrospect I am very grateful that my father was such an enthusiastic instructor and that I strived to emulate him. Being an instructor who develops learning processes with his students and who is there for them, who considers his profession more than just a job is a wonderful feeling.

9. Your father was deployed for military service at ADEBAR airship base which is located in Sanktandres (formerly Szentandrás) county. Have you ever visited the area personally? If not, do you have plans to visit there one day? Have you ever been to the Banat region?

Yes, my father was stationed at Timisoara airship base from October 1915 as an aeronaut with Königlich-Sächsischer Feldtrupp für Luftschiffe Nr. 14 (LT 14). In the past I was not familiar with Timisoara. Since writing down my father's experiences at Timisoara airship base and by connecting with Werner Kremm, reading several editions of Allgemeine Deutsche Zeitung für Rumänien (Banater Zeitung, respectively), I feel more familiar with the Timisoara region. Our Saxon newspaper released articles such as "Was ein Schwarzwälder mit Timisoara vorhat" (What are a person from the Black Forest's plans at Timisoara), or "Ein Schwarzwälder für Rumäniens Erfolgsstadt" (a Black Forest resident in Romania's successful municipality), "Ruhepol im unruhigen Rumänien" (haven of peace in turbulent Romania). I understand Timisoara to be a beautiful, interesting city, also called "Little Vienna". Time will tell if my family and I will travel to Timisoara to pay a visit to the place my father was once stationed at as an aeronaut. The freedom of travel within the European union offers a variety of opportunities to explore Timisoara and its surrounding area.

12.2 Preserving the History of Timisoara Airship Base

Within Banat local heritage societies, tradition and remembrance of cultural assets worth preserving are of high importance. This aspect of Banat culture becomes apparent in their dedication to maintain war graves. Despite how long ago these occurrences took place, for example the crash of Zeppelin LZ

Image 0-1 Above: Flyer's grave. (image: Dr. Jörg Biber)

Image 0-2 Above: Flyer's grave. (image: Dr. Jörg Biber)

Image 0-3 Above: Flyer's grave. (image: Dr. Jörg Biber)

86 near Timisoara on September 4, 1916, which killed eight aeronauts, as well as the presence of a variety of political systems in the country, its people have taken great care to maintain the graves and to remember the events. They tried to learn lessons from the occurrences, contribute to reconciliation, foster a sense of community, and to make a case for intercultural understanding.

Their extensive efforts resulted in many graves and corresponding monuments being preserved over the years, some of which had been partially renovated. Many of the graves and monuments are in exceptionally good condition, such as grave and monument facilities commemorating the aeronauts who perished in the September 4, 1916, crash of airship LZ 86 (see image 0-1 and 0-3) or the "Flieger-Grab" at Resita (see image 0-2).

The original "Flieger-Grab" "which in 1915 was dedicated to a German aircraft that had crashed near the Cosia forest house while crossing the Southern Carpathians in World War I. The pilots, Hauptmann Emil Kurt Müller (January 13, 1886-October 15, 1915) and Oberleutnant Wulfgar von Koerber (July 9, 1889-October 15, 1915) both perished in the crash. They were laid to rest at the large German cemetery in Resita; the propeller of their aircraft was used to decorate their grave." (Tigla, 2015).

The propeller was later replaced by a propeller cast at the Resita ironworks. Several German soldiers who fell in WWII were also laid to rest there.

The Banat population is very involved in the care of the gravesites. From 1992, Demokratisches Forum der Banater Berglanddeutschen (Democratic Panel of Banat Germans from the Bergland area) as well as Kultur- und Erwachsenenbildungsverein "Deutsche Vortragsreihe Reschitza" (Cultural and Continuing Education Association "German lecture series") have caring for the "Flieger-Grab". Erwin Josef Tigla initiated a fundraising campaign to benefit the renovation of the grave. Dr. Herwig Brandstetten, curator for the Österreichisches Schwarzes Kreuz (Austrian Black Cross), particularly the Steiermark headquarters who maintain war graves have generously contributed to this cause.

"Candles are lit annually on Allerseelen (All Soul's Day) and Remembrance Day in mid-November when prayer vigils are held. The German Federal consular agency in Timisoara, represented either by The Consul himself or his delegate make a pilgrimage to the Resita "Flieger-Grab" almost every year on Remembrance Day to place a wreath on the grave and pray in commemoration of the fallen soldiers from WWI and WWII." (Tigla, 2015)

On the centenary of the "Flieger-Grab", a commemoration ceremony was held at the grave site along with a documentary exhibit on the aircraft's crash which was put together by Robert Leopold Fabry and was held at the headquarters of Demokratisches Forum der Banater Berglanddeutschen.

Events for joint commemoration and grieving are also held at other grave sites. The German Consulate in Timisoara organizes a "memorial service at the 'Luftschiffer-Grab' at the military cemetery at Lippaer Strasse in Timisoara in connection with the upcoming Remembrance Day. The ceremonies are meant to commemorate the Zeppelin tragedy which took place near Timisoara on September 4, 1916, which killed multiple soldiers, as well as the combat which had reached its peak exactly 100 years ago during WWI." (ADZ, November 2, 2016). German military attaché staff representatives will also be present at the embassy in Bucharest, German panels, and local political sites.

There were also reports in the Süd-Kurier newspaper on the occasion of the Zeppelin crash in the vicinity of Timisoara. Marc Ascui who is the German military attaché representative in Bulgaria talked about the horrors of war, commemorated its victims, and referenced the September 4, 1916, Zeppelin tragedy. "A stone pillar was placed at the Timisioara military cemetery in commemoration of those who perished in the crash. In front of it are plaques with the names of each crew member.

To military official Marc Ascuri and German consul, Rolf Maruhn, the pillar serves as a reminder of the horrors of war in general." (Wagner, Südkurier, 2016).

According to them, the Zeppelin memorial pillar in Timisoara should also serve as a commemoration of the many innocent war victims in Syria, Yemen, and other seats of war world-wide.

In 1999, thanks to the efforts of Volksbundes Deutscher Kriegsgräberfürsorge (German War Graves Commission), "a group of vocational training center students from Friedrichshafen traveled to Western Romania in order to refurbish the Zeppelin memorial monument together with students from Lenau high school."

Some of the friendships that were formed back then have continued to the present day, which is very important, according to Ovideo Gant (former principal of the German speaking Nikolaus-Lenau-high school in Timisoara), 'because many people in Germany do not know what life is like in Romania.'

In that way, the tragedy surrounding the Zeppelin crash has formed the foundation for a new friendship between the two countries." (Wagner, Südkurier, 2016).

Finding solutions in a peaceful way if conflicts arose was an aspect of international understanding that was to form an integral part of commemorating the dead.

In an article posted as part of Allgemeine Deutsche Zeitung's column titled "Beiträge zur Banater Erinnerungskultur" (contributions to the Banat culture of memory), Werner Kremm wrote about the activities of airships overhead the Balkans which had departed from Timisoara (see ADZ, April 30, 2014, and ADZ, May 10, 2014).

Siegfried Thiel wrote about eight decades of Timisoara airport history titled "Am Anfang standen Zeppeline der deutschen Luftwaffe" (in the beginning, there were German Air Force Zeppelin aircraft) (ADZ, July 28, 2015). Timisoara airship base was established in 1916; it was utilized by airships, military aircraft, and balloons (particularly Drachenballone) until 1918. Following the First World War, the airship base was plundered by residents. When the airship hangar's framework was dismantled/destroyed, the hangar collapsed; five people perished in the rubble.

As the machinery house, the gas works, warehouses, and the German wireless telegraph station were raided, the possibility of future civilian flight operations was no longer an option.

This delayed the opening of the first temporary civilian airport until 1935.

Following multiple interim solutions, construction began for today's airfield in 1960. "The number of passengers traveling through Traian Vula International Airport in Timisoara increased by nearly one third in the first six months of 2015. Between January and June of that year, 427,000 passengers traveled via Timisoara airport." (ADZ, July 28, 2015).

Timisoara airship base was considered a modern airfield at the time. Since then, the new airport proved to be a worthy upgrade of the former airship base. The times of destruction and plundering of the airship base and its hangars, workshops, and plants appears to be over, turning Timisoara into an emerging, beautiful city with a stable economy and qualified professionals.

This town is certainly worth a visit.

Reprint of the article "Luftschiffhafen bei Sanktandreas" (Airship Base near Sanktandreas) from the monthly edition of Kultur- und Erwachsenenbildungsverein "Deutsche Vortragsreihe Reschitza" magazine Number 1/2017, pages 95 through 102, courtesy of author Robert Leopold Fabry

Luftschiffhafen bei Sanktandreas*

von Robert Leopold Fabry
früher Reschitza, heute Wendlingen / Deutschland

Im Jahr 1916 war der Erste Weltkrieg voll im Gange. Österreich-Ungarn und Deutschland führten unter anderen auch den Krieg gegen Rumänien. Der Grund: Rumänien hatte am 27. August 1916 Österreich-Ungarn den Krieg erklärt.

Am 29. Juli 1915 bat der deutsche Generalstabchef von Falkenhayn den k.u.k. Generalstabchef von Conrad um die Zustimmung für den Bau einer Luftschiffhalle in Südungarn (Banat). Am 18. August 1915 wurde der Vorschlag angenommen und so wurde der deutsche Feldflugchef, Major Themsen, mit dem Hallenbau nahe Temeswar beauftragt.

Der Ordonanzoffizier des Temeswarer Rayonkommandos, Oberstleutnant Graf Breda, war hiesiger Grundbesitzer und so wurde am 29. August 1915 ein Pachtvertrag zwischen der Gemeinde Neubeschenowa und der Herresverwaltung abgeschlossen. Es war die Rede von der so genannten „Hüttung über den Graben" - eine Stelle in der Größe von 900 x 400 m nahe Sanktandreas.

Um die Stelle besser zu erreichen wurden von dem Eisenbahnwächterhaus in Sanktandreas eine Schmalspurbahn und auch eine Straße gebaut.

Am 3. September 1915 wurde mit dem Bau der Luftschiffhalle begonnen. Zu diesen Luftschiffhafen gehörte eine Werkstatt, ein Vorratslager, eine Radio-, Funk- und Feldwetterstation, die Gaserzeugungslampe mit Gasflaschenschuppen und die Abfüllanlage mit Leitung. Es gab auch eine Stromerzeugungsanlage mit Umfarner und Trafostation für Blinkfeuer, Scheinwerfer, sowie einen Bomben- und Benzinkeller.

** Der Text wurde vom Autor anlässlich der Organisierung der Dokumentationsausstellung „100 Jahre seit dem Absturz des Zeppelins LZ 86 in Sanktandres bei Temeswar, am 4. September 1916" (Organisator: Robert Leopold Fabry), im Deutschen „Alexander Tietz"-Zentrum Reschitza, am 19. Oktober 2016 verfasst. Die Fotos stammen von einem Gedenken in Temeswar, vom Demokratischen Forum der Banater Berglanddeutschen und vom Kultur- und Erwachsenenbildungsverein „Deutsche Vortragsreihe Reschitza", am 21. September 2016 organisiert.*

Es waren mehrere Baracken für die Mannschaft mit Verpflegung und sogar ein Gemüsegarten für die Selbstversorgung mit frischer Ware vorgesehen. Die Gesamtkosten waren 7.554.000 Kronen. Dieser Luftschiffhafen wurde von SL 19, LZ 81, LZ 85, LZ 86 und LZ 97 angeflogen und benutzt.

Kommandanten des Luftschiffhafens waren: Hptm. Emmerich Parchner von Zobor, Hptm. Hans Kirsch, Hptm. Karl Lumbe Edler von Mallonittz, Hptm. Hermann Kühn, Hptm. Kornel Poppe, Hptm. Ferdinand Baumann - bis Kriegsende.

Ende Dezember 1917 wurde von der deutschen Heeresverwaltung dem k.u.k Kriegsministerium die Halle zum Kauf angeboten, so dass diese bis zum Ende des Krieges benutzt wurde.

Nach dem Kriegsende wurde die Halle von Privatpersonen unsachgemäß abgebaut, so dass diese am 15. Mai 1919

zusammenbrach und dabei fünf Tote und viele Verletzte forderte.

Luftschiff Zeppelin LZ 56 (LZ 86)

Dieser Zeppelin wurde in Löwental bei Friedrichshafen am Bodensee in Deutschland gebaut. Die erste Fahrt war am 10. Oktober 1915. Im Sommer 1916 wurde das Luftschiff in Dresden umgebaut (verlängert), damit es noch mehr Last tragen konnte (Bomben und Munition). Im August 1916 wurde es nach Sanktandreas verlegt und in den Dienst des Heeres gestellt.

Technische Daten:

Länge: 178,50 m
Durchmesser max.: 18,71 m
Gaszellen: 18
Volumen: 35,800 m^3
Nutzlast: 17,9 t
Steighöhe: 3200 m
Motoren: 4 Stück Maybach je 240 PS
Geschwindigkeit max.: 95,4 km/h
Reichweite: 4.900 km
Gondel: 2
Propeller: 4
Kommandanten: Linarz, Wolff

Die Ereignisse vom 4. September 1916 sind unterschiedlich dokumentiert und bewertet. Es gibt mündliche Aussagen von Zeitzeugen und Briefe seitens der Militärs an die Angehörigen der Verstorbenen.

Nach meinen Recherchen möchte ich mich hiermit nicht darauf festlegen, was richtig oder falsch ist, sondern nur versuchen zu beschreiben was passiert ist.

Robert Leopold Fabry beim Grab in Temeswar am 21. September 2016

Dabei habe ich mich an einem Brief von einem, der das Unglück überlebt hat und der die Ereignisse der Familie von einem Kameraden (Untersteuermann Karl Reiner) geschrieben hat, orientiert.

In diesem Brief steht, dass die Fahrt für 7 Uhr abends am 3. September 1916 festgesetzt wurde und dass die Besatzung aus 15 Personen, sprich 4 Offizieren und 11 Unteroffizieren bestand. Die Fahrt verlief glatt und die Aufgabe (Bombardierung) wurde glänzend erledigt. Gegen 12 Uhr mittags, am 4. September 1916 wurde der Luftschiffhafen bei Sanktandreas erreicht und für die Landung musste man Gas aus den Zellen ablassen. Folge dessen, dass das Luftschiff zu schnell von 1.400 m auf 400 m runter fiel, prallte es mit beiden Gondeln stark auf. Die hintere Gondel blieb gleich liegen und die Maschinisten wurden rausgeschleudert. Erleichtert flog das Luftschiff mit der vorderen Gondel fort. Aus cca. 80 bis 100 m fiel auch die vordere Gondel runter und die 4 Offiziere und 5 Unteroffiziere wurden unter der Gondel begraben, darunter auch der Kommandant Hptm. Walter August Wolff. Das Luftschiff ohne Gondeln mit zwei Maschinengewehr-Schützen und einem Maschinist ist schnell hoch geflogen und stürzte nach 2 Stunden in ca. 25 km Entfernung ab. Dabei brach das Gerippe in zwei Teile.

Nach drei Tagen erfolgte das Begräbnis auf dem Heldenfriedhof in Temeswar. Viele Leute und das ganze Militär nahmen an der Beerdigung teil. Die (4) Offiziere wurden in die Heimat ausgeflogen nur die (5) Unteroffiziere sind dort beerdigt. Diese sind: Ut.off. Georg Trüb, Ut.off. Jakob Renz, Ut.off. Karl Reiner, Ut.off. Otto Schienkler, Ut.off. Franz Thomsen.

Die Offiziere, die über Bukarest ausgeflogen wurden, sind: Hptm. Fritz August Walter Wolff - der Kommandant, Ltn. Heinrich Benner, Ltn. Friedrich Kraemer und Ltn. Kurt Köstlich.

Die Koffer der Gefallenen wurden mit den Habseligkeiten sowie den Wertsachen - Ringe, Uhren, Halsketten und dergleichen gepackt. Diese Koffer wurden

dem Kommando zur Beförderung mit dem nächsten Transport nach Deutschland übergeben.

Weil man im Einsatz möglicherweise sein Leben verliert, hat jeder einen Brief an die Angehörigen hinterlassen. Hier

der Text vom Abschiedsbrief Karl Wilhelms Reiner, geboren 21.02.1890 in Menden / Deutschland:

„*Viribus Unitis* (In der Kraft vereint)
Temeswar, den 3.09.1916
Liebe Eltern und Geschwister!

Solltet Ihr diesen Brief erhalten, dann lebt Eurer Bertl nicht mehr. Wir haben für die Nacht einen schweren Auftrag und erfüllt soll er werden.

Ich rechne bestimmt damit, wenn die Fahrt gemacht wird, wir müssten zurückkommen. Sollen wir in Gefangenschaft kommen, so werde ich gleich schreiben.

Lebt wohl,
Euer Sohn und Bruder, Bertl
Mein Koffer, Packet und Brieftasche wird Euch zugesandt."

Die Uhren der Gefallenen blieben auf 12 Uhr und 7 Minuten stehen!

Aktuelle Fotos:

Dokumentationsausstellung „100 Jahre seit dem Absturz des Zeppelins LZ 86 in Sanktandreas bei Temeswar, am 4. September 1916", Organisator: Robert Leopold Fabry (früher Reschitza, heute Wendlingen / Deutschland)

19. Oktober 2016, Deutsches „Alexander Tietz"-Zentrum Reschitza

Erwin Josef Ţigla und Robert Leopold Fabry

Robert Leopold Fabry spricht über die Dokumentationsausstellung, die er Dank seiner Forschungen zusammengetragen hat

The article illustrates a strong endeavor to most accurately describe specific events which occurred a century ago at Sanktandreas airship base near Timisoara. Accounts of conversations and memorial services held at the gravesites for the perished aeronauts were collected as part of a documentary exhibition.

Many of the described situations such as the construction of Timisoara airship base, information on Zeppelin airship LZ 86 (LZ 56), on its war patrol flight on September 3 and 4, as well as the burial of the fallen soldiers on September 7, 1916, were previously discussed in this book.

There are some interesting notes in this article, however, which were part of a specific preparation routine carried out by aeronauts prior to the execution of the planned war patrol mission.

As aeronauts had to expect the worst possible outcome when leaving for a war patrol mission (loss of life), each crew member wrote a letter for their family which was to be left at the airship base of departure.

Below is the farewell letter composed by Karl Wilhelms Reiner, born on March 21, 1890 in Menden/Germany:

"Virius Unitis (unified in strength)
Timisoara, September 3, 1916

Above: MG-Platform on top of a Zeppelin. (image: Schmeelke)

Dear parents and siblings,

If you should receive this letter, your Bertl is no longer living. We were tasked with a difficult mission tonight; it is important that it be fulfilled. If we launch successfully on this mission, I expect to return. Should we be taken captive, I will write to you as soon as I have the opportunity.

Farewell,
Your son and brother, Bertl
My suitcase, packet, and wallet will be sent to you."

The watches of the fallen stopped at 12:07.

The names of the fallen officers are:
Hauptmann Fritz August Walter Wolff –LZ 86 Commander
Leutnant Heinrich Brenner
Leutnant Friedrich Kraemer
Leutnant Kurt Köstlich

The belongings of the fallen LZ 86 crew members such as valuables (rings, watches, necklaces, etc.) were packed into their suitcases which were then submitted to the command responsible for transporting them back to Germany.

Above: Sound detector. (image: Schmeelke)

Above: Engine nacelle of a Zeppelin. (image: Schmeelke)

Above: Zeppelin L71. (image: Schmeelke)

Above: 20mm Becker cannon in a Zeppelin. (image: Schmeelke)

Above: Temeswar airship base. (image: Schmeelke)

Above: Zeppelin hangar at Jamboli. (image: Schmeelke)

Above: Zeppelin hall at Staaken, 1918. (image: Schmeelke)

Above: Zeppelin airship hall, Staaken. (image: Schmeelke)

Bibliography:

ADZ, 4-2014

Allgemeine Deutsche Zeitung für Rumänien (2014): **Kremm, Werner**: Beiträge zur Banater Erinnerungskultur 15 (articles in the German newspaper for Romania) / Wednesday, April 30, 2014 / www.adz.ro/artikel/artikel/beitraege-zur-banater-erinnerungskultur-15/

ADZ, 5-2014

Allgemeine Deutsche Zeitung für Rumänien (2014): **Kremm, Werner**: Beiträge zur Banater Erinnerungskultur 16 (articles in the German newspaper for Romania) / Saturday, May 10, 2014 / www.adz.ro/artikel/artikel/beitraege-zur-banater-erinnerungskultur-16/

ADZ, 7-2015

Allgemeine Deutsche Zeitung für Rumänien (2015): **Thiel, Siegfried**: Am Anfang standen Zeppeline der deutschen Luftwaffe (articles in the German newspaper for Romania) / Tuesday, July 28, 2015/ www.adz.ro/artikel/artikel/am-anfang-standen-zeppeline-der-deutschen-Luftwaffe/

Public Records Archive: Nr. 0133

Saxon state archives on Luftschiff-Trupp 14: Bestand 11348, Stellv. Generalkommando XII. Armeekorps

Biber Archives

Images and texts from Paul Biber's military service (1914-1918) as an aeronaut at Timisoara airship base as well as his role as Head of Experimental Workshop and Repair Facility at Seeflugzeugversuchskommando Warnemünde

Fabry Archive

Images and texts by Robert Leopold Fabry from Wendlingen (formerly Reschitza/Romania) covering airship LZ 56 (LZ 86) operations at Sanktandreas airship base in 1916, including the crash of LZ 56 and the burial of the fallen soliders, as well as the establishment of a gravesite and its care until the present day/ text and image release received via e-mail on August 23, 2021

Schmeelke Archive

Images of seaplanes and German airships in possession of Michael Schmeelke, Friedrichshafen/ permission for text and image use granted via email dated November 12, 2020

Belafi, 1988

Belafi, Michael (1988): Graf Ferdinand von Zeppelin. Leipzig: BSB Teubner

Bélafi und Schade, 2015

Bélafi, Michael und Schade, Thomas (2015): Der stille Helfer des Grafen Z. In: Sächsische Zeitung. 14./15. März 2015, Seite M2/M3

Biber, 2018

Biber, Jörg (2018): Luftschiffhafen Temeswar. In: FliegerRevue X 74, PPVMEDIEN GmbH, Bergkirchen, S. 10-21 und 111

Biber, 2018

Biber, Jörg (2018): Seeflugzeug-Versuchs-Kommando Warnemünde 1916 bis 1918.

In: FliegerRevue X 72, PPVMEDIEN GmbH, Bergkirchen, S. 10-34

Biber, 2019

Biber, Jörg (2019/2020): Tarnname „Adebar" – Erlebnisse von Luftschiffer Paul Biber auf dem Luftschiffhafen Szentandrás bei Temeswar 1915 und 1916. (Artikelserien)
Allgemeine Deutsche Zeitung für Rumänien (2019/2020) with Banat newspaper
BZ / February 20, 2019, page VII – part 1
BZ / February 27, 2019, page VII – part 2
BZ / March 14, 2019, page VII – part 3
BZ / March 20, 2019, page VII – part 4
BZ / March 27, 2019, page VII – part 5
BZ / April 10, 2019, page VII – part 6

Above: Zeppelin hangar at Wittmundshaven. (image: Schneeke)

BZ / April 17, 2019, page VII – part 7
BZ / April 24, 2019, page VII – part 8
BZ / May 8, 2019, page VII – part 9
BZ / May 15, 2019, page VII – part 10
BZ / May 22, 2019, page VII – part 11
BZ / July 17, 2019, page VII – part 12
BZ / July 24, 2019, page VII – part 13
BZ / July 31, 2019, page VII – part 14
BZ / August 7, 2019, page VII – part 15
BZ / August 14, 2019, page VII – part 16
BZ / August 23, 2019, page VII – part 17
BZ / August 28, 2019, page VII – part 18
BZ / September 4, 2019, page VII – part 19

Biber & Racil, 2020

Biber, Jörg and Racil, Jiri (2020): Codename "Adebar" – Erlebnisse von Luftschiffer Paul Biber auf dem Luftschiffhafen Szentendrás bei Temeswar 1915 und 1916 (Article series). Allgemeine Deutsche Zeitung für Rumänien (2019/2020) with Banat newspaper
BZ/ March 3, 2020, page VII – part 20
BZ/ March 11, 2020, page VII – part 21
BZ/ March 18, 2020, page VII – part 22

Der Bombenkrieg

Hans von Schiller, Friedrichshafen 1925: Das Zeppelin-Luftschiff im Dienst der Marine. In: Der Bombenkrieg 1914-1918, die Marineluftschiffe.
http://www.altearmee.de/luftkrieg/luftkrieg2.htm

Fabry, 2017

Fabry, Robert Leopold (2017): Luftschiffhafen bei Sanktandreas. Part of the monthly magazine Monatsschrift des Kultur- und Erwachsenenbildungsvereins "DEUTSCHE VORTRAGSREIHE RESCHITZA", number 1/2017, pages 95 through 102 / Permission for reprint granted via e-mail received on August 23, 2021

Fiedler, 1988

Fiedler, Wolfgang (1988): Fliegende Zigarren nahmen nun Kurs auf den Dresdner Hafen. In: "wir" number 42, October 21, 1988, page 6

Griehl, 2010

Griehl, Manfred (2010): Deutsche Luftschiffe seit 1871. Stuttgart, 1st edition 2010, 127 pages

Groehler, 1985

Groehler, Olaf (1985): Geschichte des Luftkrieges 1910 through 1980. Berlin: military publishing company, 6th edition 1985, 743 pages

Kornel Poppe

see: http://scharekf.hu/zeppelin.html, Kornel Poppe
Szentandrás airfield 1.
Szentandrás airfield 2.
Szentandrás airfield 3.

Kremm (2014-1):

Kremm, Werner (2014): Articles for Banat culture of memory (15), part of Allgemeine Deutsche Zeitung für Rumänien, Saturday, April 13, 2014

Kremm (2014):

Kremm, Werner (2014): Articles for Banat culture of memory (16), part of Allgemeine Deutsche Zeitung für Rumänien, Saturday, May 10, 2014

Kustan & Penz

Kustan, Alexander and Penz, Gerald: Hptm. Max Macher bei der Deutschen LuftschifferAbteilung – part 1-3 / see: https://scharekf.hu/dotroll/Hauptmann Max Macher-Teil 1-3/Hauptmann Max Macher_Arge_Feldpost_Austria_Teil 3_neu.pdf
See also: http://scharekf.hu/zeppelin.html, Max Macher

Lehmann, 2019
Luftschiffkapitän Ernst A. Lehmann (2019): Auf Luftpatrouille und Weltfahrt: Erlebnisse eines Zeppelinführers im Krieg und Frieden. epubli publishing company, 2019

Marine, 1989/IV
Marine-Luftschiffer-Kameradschaft (1989): Zeitgeschichtliche Forschungsdokumentation der ehemaligen deutschen Marine- und Heeresluftschiffahrt, book IV/1989 – permission granted for references to Fregattenkapitän a.D. Ulrich Israel

Marine, 1990/I
Marine-Luftschiffer-Kameradschaft (1990): Zeitgeschichtliche Forschungsdokumentation der ehemaligen deutschen Marine- und Heeresluftschiffahrt, Heft I/1990 – permission granted for reference to Fregattenkapitän a.D. Ulrich Israel

Marine, 1992/I
Marine-Luftschiffer-Kameradschaft (1992): Zeitgeschichtliche Forschungsdokumentation der ehemaligen deutschen Marine- und Heeresluftschiffahrt, Heft I/1992 – permission granted for reference to Fregattenkapitän a.D. Ulrich Israel

Marine, 1992/III
Marine-Luftschiffer-Kameradschaft (1992): Zeitgeschichtliche Forschungsdokumentation der ehemaligen deutschen Marine- und Heeresluftschiffahrt, Heft III/1992 – permission granted for reference to Fregattenkapitän a.D. Ulrich Israel

Marine, 1992/IV
Marine-Luftschiffer-Kameradschaft (1992): Zeitgeschichtliche Forschungsdokumentation der ehemaligen deutschen Marine- und Heeresluftschiffahrt, Heft IV/1992 – permission granted for references to Fregattenkapitän a.D. Ulrich Israel

Mückler, 2013
Mückler, Jörg (2013): *Deutsche Flugzeuge im Ersten Weltkrieg*. Stuttgart, 1st edition 2013, 223 pages

NDR.de, 04.11.2016
Gressieker, Oliver (2016): *Wittmundhafen: Vom Zeppelin bis zum Eurofighter*. NDR.de – Geschichte-Orte. 04.11.2016, 8.08 Uh

Reprint of 1920
Die gesamten deutschen Luftstreitkräfte im 1. Weltkrieg. 1920, Springer Verlag, reprint 2011, Europäischer Hochschulverlag GmbH & CoKg Bremen

Nelepcu, 2020
Raluca Nelepcu from Timisoara, editor at Allgemeine Deutsche Zeitung für Rumänien and translator (Romanian-German/Romanian-English)

Ploes, 2018
Ploes, Hans-Günther (2018): Zeppeline über dem Balkan – Temesvár-Szentandrás 1915-1917. In: Luftwaffen-Revue. Nr. 1/March 2018, NETTEVERLAG; Nettetal; pages 16-23

Post LZ 81
Postgeschichte des Heeresluftschiffs LZ 81 (LZ 51)
www.ezeptalk.de/Das Forum für Zeppelinpost
Image 110:
also www.veikkos.com/bilder/5/luftschiff_zeppelin_Kommando_Lz_81_w0357315.jpg
Image 218
also: https://commons.wikimedia.org/wiki/file:Siegelmarke_Kommando_Zeppelin_
Luftschiff_L.Z_97

Redner
Redner, Harry C.: Die Luftschiffwaffe des Heeres/Die Geschichte der deutschen Heeresluftschiffahrt. http://www.luftschiffharry.de

SANKTANDRES, 1981
Weber, Matthias und Dr. Anton P. Petri (1981): SANKTANDRES – Heimatbuch (hometown book) Sanktandres in the Banat region, including numerous images and sketches. Published by

Heimatortsgemeinschaft
Sanktandres, 1981

Sammlung Racil
Jiri Racil from Brno, Czech historian, see www.hi-story.cz and www.oldphoto.info

scharekf.hu
Image 5 /dotroll/besenyo_elemei/Poppe-S2-14-k.jpg
Image 6 /dotroll/besenyo_elemei/Poppe-S2-26-k.jpg
Image 7 /dotroll/besenyo_elemei/tn_/image09.jpg
Image 8 /dotroll/besenyo_elemei/image028.jpg - Siegelmarke
Image 60 /dotroll/besenyo_elemei/image026.jpg
Image 61 /dotroll/besenyo2_elemei/image010.jpg
Image 74 /dotroll/besenyo_elemei/Poppe-S2-26-k.jpg
Image 77 /dotroll/besenyo_elemei/image021.jpg
Image 78 /dotroll/besenyo_elemei/image022.jpg
Image 92 /dotroll/besenyo3_elemei/LZ_81_Barogramm-8-11-1915_Sofia_Los-1255_k.jpg
Image 93 /dotroll/besenyo3_elemei/LZ_81_Sofia_Flugfeld_1.jpg
Image 117 /dotroll/besenyo3_elemei/image013.jpg
Image 146 /dotroll/besenyo3_elemei/image013.jpg
Image 173 /dotroll/besenyo2_elemei/image015.jpg
Image 174 /dotroll/besenyo3_elemei/LZ85__Motorengondel:Temesvar_1.jpg
Image 176 /dotroll/besenyo3_elemei/image006
Image 184 /dotroll/besenyo_elemei/image007.jpg
Image 200 /dotroll/besenyo_elemei/imagexx.jpg
Image 201 /dotroll/besenyo3_elemei/image051.jpg
Image 206 /dotroll/besenyo2_elemei/image019.jpg
Image 207 /dotroll/besenyo2_elemei/image007.jpg
Image 211 /dotroll/besenyo2_elemei/image016.jpg
Image 214 /dotroll/besenyo3_elemei/magasságmérő.jpg
Image 216 /dotroll/besenyo3_elemei/LZ_81_Bruchlandung_in Tirnova_[Bulgarien]_1.jpg
Image 217 /dotroll/besenyo3_elemei/LZ_97_Besatzung-Szentandras_1.jpg
Image 219 /dotroll/besenyo3_elemei/image069.jpg

Schneevogt und Schulz (2007)
Schneevogt, Jürgen und Schulz, Heinz (2007): *Die militärische Nutzung der sächsischen LuftschiffHäfen Dresdun und Leipzig 1912–1919.* Strausberg 2007, 78 Seiten

Schwaebische Zeitung (newspaper), 2015
Friederica Ihling (November 14, 2015): Luftschiffe im Einsatz über dem Balkan.

Stadtwiki Dresden
Luftschiffe in Dresden. / see http://www.stadtwiki.de/wiki/Luftschiffe

SZ, 2010
Sächsische Zeitung (August 31, 2010): Luftschiffe kehren nach hundert Jahren nach Dresden zurück. SZ, page 19

Täger, 2009
Täger, Hannes, Dr. (2009):
Friedens- und Kriegsfahrten des LZ 17 "Sachsen". In: DAS PROPELLERBLATT, Number 25, Sommer 2009, pages 899–920

Thiel (2015)
Thiel, Siegfried (2015): *Am Anfang standen Zeppeline der deutschen Luftwaffe. In: Allgemeine Deutsche Zeitung für Rumänien,* Dienstag, 28 July 2016.

Tigla (2015)
Tigla, Erwin Josef (2015): 100 Jahre "Flieger"-Grab in Reschitza. In: Monatsschrift des Kultur- und Erwachsenenbildungsverins *"Deutsche Vortragsreihe Reschitza"*, Nr. 12/2025, S.11

VDI 61.1917
VDI-Zeitschrift 61.1917 S. 485-486, Biographien bedeutender Techniker, Ingenieure und Technikwissenschaftler. 2nd edition. Berlin 1987, pages 249–257

Wagner (2016)
Wagner, Thomas (2016): Gedenkfeier an Zeppelin-Abstrutz in Westrumänien. In: Süd-Kurier vom 6. November 2016.

WIKIPEDIA Dudestii Noi
(Dudestii Noi = German Neubeschenowa)
LZ 85 overhead the Neubeschenowa airship hangar-downloadable version
mons.wikimedia.org/w/index.php?cuvid=43166794

ZEPPELIN MUSEUM BLOG, 2017
Friederica Ihling (2017): Rekordfahrt nach Süden – Luftschiff L 59 auf dem Weg nach Afrika. MUSEUMSBLOG, 24 November 2017
https://blog.zeppelin-museum.de/2017/11/24/afrikafahrt-l-59/

X1
NDR.de, 04.11.2016
Gressieker, Oliver (2016): *Wittmundhafen: Vom Zeppelin bis zum Eurofighter.* NDR.de – Geschichte-Orte. 04.11.2016, 8.08 Uh

X2
Schneevogt und Schulz (2007)
Schneevogt, Jürgen und Schulz, Heinz (2007): *Die militärische Nutzung der sächsischen Luftschiffhäfen Dresden und Leipzig 1912–1919.* Strausberg 2007, 78 Seiten

X3
Thiel (2015)
Thiel, Siegfried (2015): *Am Anfang standen Zeppeline der deutschen Luftwaffe. In: Allgemeine Deutsche Zeitung für Rumänien*, Dienstag, 28. Juli 2016.

X4
Tigla (2015)
Tigla, Erwin Josef (2015): 100 Jahre "Flieger"-Grab in Reschitza. In: Monatsschrift des Kultur- und Erwachsenenbildungsvereins "*DEUTSCHE VORTRAGSREIHE RESCHITZA*", Nr. 12/2015, S. 11

X5
Wagner (2016)
Wagner, Thomas (2016): Gedenkfeier an Zeppelin-Absturz in Westrumänien. In: Süd-Kurier vom 6. November 2016

Made in United States
Orlando, FL
04 November 2026

90200871R00117